STYLE SHANGHAI

2015/2016秋冬海派时尚流行趋势
2015/2016 AUTUMN/WINTER STYLE SHANGHAI FASHION TREND

海派时尚流行趋势研究中心 著

东华大学出版社

编辑委员会

名誉主任：厉无畏

顾问：徐明稚、Christopher Breward、李耀新、席时平、邵隆图

主任：刘春红

执行副主任：刘晓刚

副主任：陈跃华、贺寿昌、解冬、朱勇

委员：林艺、庄培、吕苏宁、邵峰、蒋智威、卞向阳、徐晶

创作团队：李峻、曹霄洁、顾雯、吴晨荣、罗竞杰、沈沉、倪明、俞英、田玉晶、傅婷、吴亮、赵蔚、段然、蔡凌霄、唐强、厉莉、边菲

采编助理：傅白璐、李芷玥、崔彦、李超逸、杨楷浪、聂焕玲、梅伶俐、高宁馨

主题摄影：TO2X 罗竞杰

插画：胡娱曼

支持单位：

上海市经济和信息化委员会

上海市文化创意产业推进领导小组办公室

上海市长宁区人民政府

东华大学

上海智富企业发展（集团）有限公司

司色艾印染科技（上海）有限公司

上海国际时尚联合会

上海国际服装服饰中心

上海服饰学会

依托平台：

环东华时尚创意产业服务平台

海派时尚创意与设计知识服务平台

基于云计算的创意设计公共服务基地

目录 CONTENTS

序 PREFACE	4
2015/2016 秋冬海派时尚流行趋势主题：革新·生机 2015/2016 AUTUMN/WINTER STYLE SHANGHAI FASHION TREND THEMES: RENOVATION·VITALITY	6
专家观点 EXPERTS' VIEWPOINTS	8
邵隆图 Shao Longtu	8
刘晓刚 Liu Xiaogang	10
卞向阳 Bian Xiangyang	12
潘德蒙 Ron Pedemonte	14
社会背景 SOCIAL BACKGROUNDS	16
政治 POLITICS	16
经济 ECONOMY	17
科技 TECHNOLOGY	18
环境 ENVIRONMENT	19
生活 LIFESTYLE	20
艺术 ART	21
趋势演绎 TREND INFERENCE	22
塑 CREAT [海派经典风格 SHANGHAI CLASSIC STYLE]	24
主题说明 TREND THEME INFERENCE	26
主题色彩 COLOR THEME	29
中西合璧的海派繁华 STYLE SHANGHAI PROSPEROUS WITH ELEGANT SMART	30
时装设计中的东方哲学 ORIENTAL PHILOSOPHY IN FASHION DESIGN	32
以年轻之心传承海派首饰 HERITAGE OF SHANGHAI JEWERY WITH YOUNG HEART	34
精致帽饰中的中国元素新演绎 NEW CHINESE ELEMENTS IN ELEGANT HAT DESIGN	36
传统手工的传承与发扬 THE INHERITANCE AND PROMOTION OF THE TRADITIONAL HANDWORK	38
传统形象的时尚演绎 FASHION INTERPRETATION OF TRADITIONAL IMAGE	40
海派文化符号的现代应用 THE MODERN APPLICATIONS OF SHANGHAI STYLE CULTURE SYMBOL	42
厚重中凝练时尚之美 AESTHETIC OF STYLE IN STRENGTH	44

2015/2016秋冬海派时尚流行趋势

塑造有意味的形式	SHAPE THE MEANINGFUL FORM	46
现代与传统相结合的配饰设计	A COMBINATION OF MODERN AND TRADITIONAL ACCESSORIES DESIGN	48
续 SUSTAIN	[海派自然风格 SHANGHAI NATURAL STYLE]	50
主题说明	TREND THEME INFERENCE	52
主题色彩	COLOR THEME	55
记忆中的江南	MEMORY OF JIANGNAN	56
传统水墨的当代性与国际化	THE CONTEMPORARINESS AND INTERNATIONALIZATION OF TRADITIONAL INK PAINTING	58
欧化简约与海派风韵	SIMPLIFIED WESTERN STYLE SHANGHAI	60
设计"真生活"	DESIGN "TRUE LIFE"	62
当建筑邂逅绿色时尚	WHILE ARCHITECT MEETS GREEN FASHION	64
有机形态演绎自然美感	INTERPRETATION OF NATURAL BEAUTY BY ORGANIC FORMS	66
蔓延于自然的时尚憧憬	EXPECTATION OF STYLE IN NATURE	68
时尚面料设计中的可持续主题	SUSTAINABLE FASHION THEME IN FABRIC DESIGN	70
创新水墨画的追求	PURSUIT OF INK PAINTING	72
首饰留下永恒自然美好	ETERNAL BEAUTY BY JEWELRY	74
趣 FUN	[海派都市风格 SHANGHAI URBAN STYLE]	76
主题说明	TREND THEME INFERENCE	78
主题色彩	COLOR THEME	81
快乐的摩登东方时尚	MODERN ORIENTAL FASHION WITH HAPPINESS	82
设计的跨文化沟通与趣味	THE CROSS-CULTURAL COMMUNICATION AND TASTE OF THE DESIGN	84
平面设计的新媒体进化	THE NEW MEDIA REVOLUTION OF GRAPHIC DESIGN	86
妙趣横生老上海	THE OLD SHANGHAI FULL OF WIT AND HUMOR	88
个性设计重塑，创造与乐活并行	THE REMODELING OF THE INDIVIDUAL DESIGN	90
箱包设计的年轻化、多彩化趋势	YOUNG AND COLORFUL DEVELOPMENT FOR BAG DESIGN	92
在矛盾中感受设计趣味	TO EXPERIENCE THE INTERESTING OF THE DESIGN IN THE CONTRADICTION	94
趣味性是返璞归真的艺术写照	INTERESTING DESIGN IS BACK TO BASICS	96
玩乐主义风潮下的小首饰	A HIGHLIGHT OF LIFE ATTITUDE BY FASHION DESIGN OF JEWELRY	98
在玩味中寻求释放	RELEASE IN FUN	100
"趣"看上海	INTERESTING LOOK AT SHANGHAI	102

变 VARIETY	[海派未来风格 SHANGHAI FUTURISTIC STYLE]	104
主题说明	TREND THEME INFERENCE	106
主题色彩	COLOR THEME	109
智能物联的时尚	THE FASHION OF IOT	110
亚洲灵感创造未来商业价值	THE ASIAN INSPIRATION CREATES THE FUTURE BUSINESS VALUE	112
设计是提炼日常生活的状态	DESIGN IS REFINED FROM EVERYDAY LIFE	114
简单的新鲜	SIMPLE FRESHNESS	116
闪耀的未来主义设计	SPARKLING FUTURISTIC DESIGN	118
未来帽饰中的感性创作	EMOTIONAL CREATION IN FUTURE HATS	120
年轻、自信与自由实验	THE YOUTH, SELF-CONFIDENCE AND FREEDOM TO EXPERIMENT	122
颓废中重生的新科技艺术	REBIRTH OF NEW TECHNOLOGY ART IN DECADENCE	124
"不变"中的变化	THE CHANGE IN "INVARIABILITY"	126
畅想科技与太空	THE PAMPER IMAGINATION OF THE TECHNOLOGY AND SPACE	128
未来配饰设计的建筑灵感	THE ARCHITECTURAL INSPIRATION OF THE FUTURE ACCESSORIES DESIGN	130
海派未来风格的变革思辨	ANALYZE THE REVOLUTION OF NEW SHANGHAI STYLE	132
鸣谢 ACKNOWLEDGEMENT		134

序

  2015/2016秋冬海派时尚流行趋势以"革新•生机"为主题，从经典、自然、都市、未来四大艺术风格阐述了具有本土特色的流行趋势。其中，海派经典风格演绎了对传统文化的重新演绎设计与革新，海派自然风格描述了尊重自然的可持续设计的革新，海派都市风格反映了散发都市年轻正能量的创造革新，海派未来风格带来了提升生活便捷的时尚高科技革新。

  《海派时尚》出版一年来，得到了国内外社会各界的广泛关注，这不仅是对研究的鼓励和支持，更是对发扬和推广本土海派时尚文化的支持。上海是联合国教科文组织创意城市网络"设计之都"，秉承"海纳百川、追求卓越、开明睿智、大气谦和"的上海城市精神，本书期待带给作者精彩的海派时尚故事与设计灵感，让海派文化成为时尚创意设计之源头。

  为了生动地展现设计流行趋势，本书汇集了几十名海派时尚艺术家、设计师和品牌的作品和观点，以跨界的形式共同构成流行趋势的灵感故事。以此为基础，2015/2016秋冬海派时尚色彩、面料、男装、女装、鞋履、箱包、首饰、陈列等流行趋势将在上海设计之都公共服务平台"海派时尚"（www.style.sh.cn）中陆续刊登。

  本书得到了上海市、长宁区、东华大学各级领导和社会各界的广泛支持。团队以东华大学设计学科专家为主体，汇集了艺术家、设计师、企业家、国际友人，团队成员近百名，共同迸发创意火花，数易其稿，历时半年完成。2014年初，针对本期流行趋势，特别举办了"海派时尚－创意发声"海派时尚创意大师跨界沙龙，多名嘉宾的精彩观点成为本书的创意来源。

  时尚设计产业的发展动力在于时尚文化，而时尚文化的根源在于流行趋势。衷心希望《海派时尚》对打造具有海派特色的时尚产业有所贡献。

海派时尚流行趋势研究中心

海派时尚官方网站
www.style.sh.cn

海派时尚官方微信

PREFACE

The Shanghai Style Fashion Trends of 2015/2016 Autumn and Winter interpreted the fashion trend with local characteristics from the four artistic styles, i.e., classic, nature, city, future, based on the theme of "Revolution · Vitality". Among them, the classic Shanghai Style demonstrated the re-interpretation of design and innovation for the traditional culture; the natural Shanghai Style described the sustainable design innovation of the respect for the nature; the urban Shanghai Style reflected the creation and innovation expressed with the urban young positive energy; the future Shanghai Style brought the stylish high-tech innovation enhancing the convenience of the life.

It has been widely concerned by the people from different social walks at home and abroad since STYLE SHANGHAI was founded a year ago, which is not only the encouragement and support for this book but also the support for the development and promotion of the local Shanghai Style Fashion culture. Shanghai is "the city of design" of the UNESCO Creative Cities Network and it adheres to the Shanghai spirit of "be tolerant to diversity; in pursuit of excellence; be enlightened and sagacious; be humble and powerful ". This book looks forward to bringing the readers with wonderful Shanghai Style Fashion stories and design inspiration to make the Shanghai Style Culture become the source of creative fashion design.

To give a more vivid demonstration of the design trend, this book collects the works and opinions of dozens of Shanghai Style Fashion artists, designers and brands to form the inspired stories of the fashion trend in the form of cross-border. The Shanghai Style Fashion colors, fabrics, men's wear, women's wear, shoes, bags, jewelry, and other fashion trend of the 2015/2016 autumn and winter will be gradually published on the public service platform of the design city of Shanghai, STYLE SHANGHAI (www.style.sh.cn).

The book has been widely supported by the leaders of Shanghai City, Changning District, Donghua University and people from all walks of the society. The main bodies of the team are the experts in the design disciplines of Donghua University with a collection of artists, designers, entrepreneurs and international friends, amounting to near 100 team members. It takes them six months to complete this book with the common burst of creative sparks and repeated revisions of the draft.

The development power of the fashion design industry is in the fashion culture and the origin of the fashion culture is rooted in the fashion trend. We sincerely hope that the STYLE SHANGHAI will make some contribution to building the creative fashion design industry with characteristics of Shanghai city.

Style Shanghai Fashion Trend Research Center

www.style.sh.cn   wechat code

2015/2016秋冬海派时尚流行趋势

2015/2016 AUTUMN/WINTER STYLE SHANGHAI FASHION TREND

2015/2016秋冬海派时尚流行趋势主题
2015/2016 AUTUMN/WINTER STYLE SHANGHAI FASHION TREND THEMES

革新·生机
RENOVATION·VITALITY

2015/2016秋冬海派时尚流行趋势

专家观点 EXPERTS' VIEWPOINTS

邵隆图

Shao Longtu

九木传盛品牌推进机构董事长
上海市品牌建设工作联席会议专家委员会委员
上海国际时尚联合会副会长
上海创意设计中心首席策划大师
上海创意产业中心专家委员会副主任
东华大学服装·艺术设计学院兼职教授

今天的世界，是个创意、技术、专利无限创造的时代，也是文化迅速发展更新的时代。世上本无东南西北之分，亦无一到无穷大之区别，一切皆是人为设定的标准，是不同的文化将其界定并区分。这就是创新！

上海，古属吴越，曾是远东第一时尚与金融中心，拥有深厚的近代城市文化底蕴和众多历史古迹，江南的吴越传统文化与各地移民带入的多样文化相融合，形成了特有的海派文化。上海具备海派时尚的都市基因，在建设"设计之都"的热潮下，需要充分探究海派时尚的文化基因，在发挥新时代优势的同时，更需要传承经典文化。

"上善若水，海纳百川"是上海的城市精神，海派时尚的创意源泉就是"水文化"的创新。如同没有固定形态的水，时而波澜壮阔；时而微波荡漾；亦可涓涓细流；亦可灵动清秀。海派时尚需要凭借创造力与想象力，将时尚创意包融一切。她既是东方也是西方；既是本土也是国际；既是经典也是时尚；每天都以新鲜的形态出现。

在富裕而浮躁的当下，时尚不仅意味着表面的鲜活，我们更需要传承与革新，赋予本土文化以新的内涵，也就是有我们自己文化特色的软实力。

如同海派文学的代表人物刘半农在留洋期间发明的中文文字中的"她"，海派时尚充满了女性化的阴柔，有如同母亲般开阔的胸怀、关爱与包容。

上海，不仅仅要有力度，更要有风度；
上海，不仅仅要有速度，更要有精度；
上海，不仅仅要有高度，更要有气度；
上海，不仅仅要有广度，更要有深度。

具备又嗲又嗀(jia)气质的海派时尚，正成为上海城市的表情和内涵。

她的胸怀，她的教养，她的典雅，她的智慧，她的微笑。

2015/2016秋冬海派时尚流行趋势是对经典的传承，传统的再造，自然的呵护，个性的主张。这里充满了革新，充满了生机！

2015/2016 AUTUMN/WINTER STYLE SHANGHAI FASHION TREND

Experts' Viewpoints

The world today is an infinitely creative era of ideas, technologies and patents and it is also a rapid development era of culture regeneration. There wouldn't be any differences for four cardinal points or one to the infinity in the world. Everything is artificially set to be a standard and it is defined and distinguished by the different cultures. This is innovation!

In the old times, as a part of the Wuyue Kingdom, Shanghai was the first fashion and finance center in the far east, as it has the profound modern urban culture and numerous historical monuments, integrating the traditional culture of Wuyue in the South of the Yangtze River with the diverse cultures brought by the immigrants from different places to form the unique Shanghai Style Culture. Shanghai has the urban genes for the Shanghai Style Fashion and under the booming of constructing "the city of design", and the culture genes of its fashion needs to be fully explored. Furthermore, it is more required to inherit the classic cultures while giving full play to the superiority to the new era.

"The good is like to be tolerant to diversity" is Shanghai's city spirit and the creative source of the Shanghai Style Fashion is the innovation of the "water culture". As there is no fixed form of water, sometimes it is magnificent; sometimes it is rippling; it may also be trickling, agile or delicate. The Shanghai Style Fashion needs to incorporate the fashionable originality into everything by way of the creativity and imagination. She is both in the western and eastern style; she is both in the domestic and overseas fashion; she is both in classic and stylish; she is presented with new forms and shapes every day.

In the rich and blundering times, fashion does not only mean the fresh and vivid surface, but what we need to do is the inheritance and innovation, giving new meanings to the local culture, which means that we should have the soft power of our own cultural characteristics.

Just like the "she" in the word invented by Liu Bannong, the representative of Shanghai Style Culture, when he was studying abroad, the Shanghai Style Fashion is full of the feminization of the feminine, as if mother-like open-minded, caring and tolerant.

Shanghai shall not only have the strengths but also have the demeanor;
Shanghai shall not only have the speed but also have the precision;
Shanghai shall not only have the height but also have the grace;
Shanghai shall not only have the breadth but also have the depth.

The affectedly sweet Shanghai Style Fashion with temperament is gradually becoming the city's expression and connotation.

Her mind, her upbringing, her collections, her wisdom and her smile.

The 2015/2016 autumn and winter trends of Shanghai Style Fashion is a heritage to the classic, a recycling to the tradition, a care to the nature and an assertion to the personality. Here it is full of innovation and full of vitality!

2015/2016秋冬海派时尚流行趋势

专家观点
EXPERTS' VIEWPOINTS

刘晓刚
Liu Xiaogang

教授，博士生导师
全国十佳服装设计师
东华大学服装·艺术设计学院副院长

文化崛起、大国梦想、单独多育……向往幸福的百姓必定盼望幸福。

物价指数、就业环境、周边军情……不太轻松的民众极欲追寻轻松。

凡此种种，或潜移默化，或激情呈现，将不可避免地成为影响社会形态变化的新因素，并将挟裹人们生活方式发生与之呼应的改变。作为创导生活方式的时尚流行趋势，必须研究新的社会变化的风向标。当这些因素波及时尚流行趋势，又将成为众望所归的融经典、自然、都市、未来于一体的物化产品。因而，新一轮海派时尚流行将以此为基点展开。

社会文化的潮涌、经济转型的大势、海派文化的特征决定了本季海派时尚充满着变化，民众对幸福指数的企盼呼唤着时尚的"革新"。"革新"带来"生机"，两者交织在一起，成为本季流行趋势的核心关键词。

本季的"革新"是有度的"新"，有度则美，无度则乱。稳中谋变、适度求新，仍是本季海派时尚主流。预测、猜想、创意，所有的"革新"将在有度的前提下展开，本季趋势是所有时尚人对"度"的准确拿捏。

本季的"生机"是"革新"带来的衍生，体现为民众对"轻松"生活方式的追求，旨在缓解源自各个层面的压力。廓型、色彩、材料、功能，甚至价格，无不显现出人们对这一趋势的希翼。

2015/2016 AUTUMN/WINTER STYLE SHANGHAI FASHION TREND

专家观点
Experts' Viewpoints

The rise of culture, dream of being a big country, separate, prolific......People aspiring to the happiness will surely hope for the happiness;

Price index, employment environment, surrounding military information ......People who are less relaxed are eagerly in pursuant of the relaxation.

All these, or subtle or passionate presentation, will inevitably become the new factors affecting the social morphological changes and make people's lifestyle change with them. As the fashion trends leading people's lifestyles, the new barometer of social changes must be studied. When these factors affect fashion trends, they will be the populated materialized products integrated with the classic, nature, city and the future. Thus the new generation of the Shanghai Style Fashion trend will be carried out based on it.

The social culture tide, the economic restructuring trend and the characteristics of the Shanghai Style Culture decide that this seasonal fashion is full of changes and people's hopes for happiness index call for the "innovation" in fashion. The "innovation" is to bring the "vitality" and these two words are intertwined together to be the key words of this season's fashion trends.

This season's "innovation" is an "innovation" to some extent. It will be beautiful with the extent and out of order without it. The stability and change and proper innovation are still the mainstream fashion in this season. All of the "innovations" of prediction, guess and creation will be carried out under the premise of some extent and this seasonal trend is the accurate holding of the extent made by all the fashion people.

This season's "vitality" is the derivative brought by the "innovation", which is embodied in the way that people are in pursuant of a "easy" way to live their lives aiming at alleviating the pressure from all levels. Everything from the profile shape, color, material, function, or even the price reveals that people are hoping for this trend.

2015/2016秋冬海派时尚流行趋势

专家观点 EXPERTS' VIEWPOINTS

卞向阳
Bian Xiangyang

教授、博士生导师
上海纺织服饰博物馆馆长
东华大学服装·艺术设计学院艺术学理论部主任

　　所谓"趋势",是我们对于未来理性的思辨,它以海派时尚的传承和发展为主线,建立在历史记忆的基础之上,立足现在,寻求未来的生机前程。

　　所谓"海派",意味着追欧楫美、包容四方、创新发展,它以自由贸易、商业社会和契约精神为背景,通过形式和内容的变革创新而独树一帜。

　　变革是海派时尚的本质,从时髦、摩登到简练、开放乃至21世纪的国际化,上海的时尚因为不断变革而形成了自己的风格,并受到中国乃至世界的仿效和关注。

　　生机是海派时尚的追求,上海的时尚一直在本土与西方的文明冲撞与交融中制造发展机遇,民众由此有了流行装扮,产业因此得到诸多商机。

　　海派时尚是一个活化的生态系统,它以上海的城市生活为背景,是物质与文化互为表里的体系循环。海派流行服饰的繁华与繁荣,背后不仅有时尚产业的支撑,也是海派文化的物化体现。海派时尚是知行合一的综合体,20世纪30年代的上海人就已经感觉到,摩登不是简单的烫发红唇、旗袍时装、跳舞交际,最为重要的是要有一颗Modern的心。

　　流行趋势预测是现代时尚产业的重要环节,其构建在对于产业走向和民众欲望的综合理解、判断的基础之上,着重表述未来某一可以预见的时段中的流行主题和表现。它具有理性的内核和感性的表述,并赋予观者较大的想象空间,意在提前影响民众的时尚思维,引导时尚产业的资源聚合,通过潜移默化,最终形成海派时尚在产业和民众中的话语权。

　　张爱玲曾经有言:"上海的裁缝是没有主张的",那是因为上海人有天生超前的时尚意识与品味。今后的海派时尚,是民众和产业一起在变革中求生机,共同实现美丽的中国梦想。

专家观点
Experts' Viewpoints

2015/2016 AUTUMN/WINTER STYLE SHANGHAI FASHION TREND

The "trend" is defined as our rational thinking for the future, with its heritage and development of Shanghai Style Fashion as main clue, based on the historical memory, present life and the future vitality prospect.

The "Shanghai Style" means the pursuit to Europe and United States, inclusiveness, innovation and development, with free trade, business community and spirit of contract as the background, and flying its own colors through the change and innovation of form and content.

Change is the essence of Shanghai Style. From being stylish and modern to concise and open, and even international in this century, Shanghai's fashion forms its own style due to constant changes, followed and focused in China and the world.

Vitality is the pursuit of Shanghai Style. Shanghai's fashion has been creating the development opportunities in the course of collision and fusion between Chinese and Western cultures, so that the people have obtained the popular dressing, and there are a lot of business opportunities in the industry.

Shanghai Style Fashion is an activation of ecosystem, and the circulative system of material and cultural interacting in the context of Shanghai city life. The bustling prosperity of fashion in Shanghai reflects the support of fashion industry and also represents the materialized Shanghai culture. Shanghai Style Fashion is a complex of knowledge and action. The Shanghai people in the 1930s had already felt that the modern is not simply perm and red lip, cheongsam fashion and dance, the most important thing is to have a modern heart.

Trend forecasting is an important part of modern fashion industry, which is built on the comprehensive understanding and judgment of industry trends and public desires, focusing on the presentation of popular theme and performance in a foreseeable future period. It has the rational kernel and emotional expression, and gives a greater imagination to the viewers, with the purpose of influencing the people's fashion thinking, guiding the resource integration of fashion industry, and ultimately forming the discourse right of Shanghai Style Fashion in the industry and the public through subtle influence.

Eileen Chang once said that "Shanghai tailors do not have their own ideas", because the Shanghai people have the natural and advanced fashion awareness and taste. In the future, Shanghai Style Fashion will be a platform for the people and industry to together survive in the transformation and to achieve a beautiful Chinese dream.

2015/2016秋冬海派时尚流行趋势

专家观点 EXPERTS' VIEWPOINTS

潘德蒙
Ron Pedemonte

康涅狄格大学博士
University of Connecticut / Doctor
Americas/Color Solutions International / 司色艾 总裁

从艺术到科学——将你看到的色彩实现并应用。

如何在每一季色板中实现正确的颜色？色彩沟通和管理将支持和帮助你迅速有效地适应当今"顾客"为主导的市场。

沟通中最重要的一步是颜色必须精准地传达。我们认为，当今设计本身的步骤不需要改变，必须改变的是如何精确的传达设计灵感。

全球采购的发展，让我们发现在供应链过程中，错误的色彩沟通在不断地重复发生，色彩的牢度及性能达不到要求；色彩没有跟上当季的时尚流行趋势而导致交期延误；或者色彩与设计师原先的灵感不一致。色彩的问题被认为是产品创新进步的最大因素。

这就需要有一个精确的、科学的、能够将设计灵感准确表达的方式方法。一个好的颜色标准背后的隐含价值是它将引领一个公司在走向市场的过程中避免一些可预见的失误以及使客户对这个品牌更忠诚，并且让这个品牌更有特点。

有效的颜色管理是快速的色彩分发模式，这个模式使品牌或零售商的供应商快速地得到这个颜色。

时尚快速的发展转换，新的设计不断推出，这就需要一个快速的色彩沟通过程。电子颜色标准（QTX文件）应运而生，它是颜色的一组光谱数据或者说是色彩的"指纹"，是通过分光光度计的口径测试产生的物理反应。这组电子数据可以让工厂在拿到实物颜色之前或实物颜色在运送过程中就提前做实验室打色，从而节省时间。

建立有效的色彩管理系统对于流行趋势的应用是必不可少的。色彩的转达和沟通将贯穿整个供应链，让色彩更为便捷地从艺术转为科学。

From Art to Science -- Helping You See What You See

So how can you achieve the right colors each season? The color communication & management support you and help you respond quickly and effectively to today's "shopper" driven market.

One of the most important factors is that the color must be communicated accurately. In our approach, the design process essentially does not change; but what does change is how the inspirational color is communicated.

Due to global sourcing, the miscommunication of color has increased rapidly, especially throughout the supply chain process. Garments that fail industry fastness and performance requirements, miss a current fashion trend due to late deliveries, or don't match the original inspiration of the designer are just a few of the problems that have surfaced. Color is identified as the largest factor impeding real improvements.

The solution to accurately communicate the color is to provide a scientific translation of the color called the "color standard". The properties of the color standard are the "substance" that prevents companies from making costly mistakes and consumers to remain loyal to the brand or label.

Fast fashion has shortened the time lines to introduce new designs, and in order to keep pace, the color communication process must be faster. Today's best practice for communicating color is to use both the physical and digital color standard (also known as the Qtx-file). The digital color standard is the spectral data or "thumbprint" of the color. It is generated on a calibrated spectrophotometer and provides the manufacturer with an accurate, electronic standard that enables the mill to immediately begin the lab dipping process.

Building an efficient color management system for your company is a work in progress. An efficient color management system must seamlessly distribute both the physical and electronic "color standard" from the Brand or Retailer to their supply chain. The color choices can be translated and communicated throughout the supply chain, and allow an easy transition of color from "art" to "science".

2015/2016秋冬海派时尚流行趋势

社会背景 SOCIAL BACKGROUNDS

政治 | POLITICS

中国开启全面深化改革大序幕，新蓝图细描美丽中国梦；

国家治理体系和治理能力现代化着力推进；

上海作改革开放排头兵，政府职能转变气象一新；

世界格局处于大发展与大变革时期，国际环境复杂多变。

China opens a prelude to the reform and the new blueprint delicately describes the beautiful Chinese dream;

Focus on promoting the modernization of the national governance systems and governance capacity;

Shanghai is considered to be the vanguard of reform and opening up, the transformation of government functions seems to be brand new;

The world situation is in the period of great development and great changes and the international environment is complex and changeable.

经济 | ECONOMY

互联网金融、4G通讯、电商，最活跃的产业群体，撬动经济发展大杠杆；

调降GDP增速，绿色经济增强百姓幸福指数；

上海聚焦制度创新，中国上海自由贸易试验区建设一步一个脚印；

中国经济运行稳中向好，结构调整稳中有进，转型升级稳中提质；

世界经济复苏形势不稳定，经济格局深度调整，国际竞争更趋激烈。

Internet banking, 4G communication, e-commerce, the most active industry groups leverage the big lever of the economic development;
Lower down the speed of GDP growth and the green economy enhances the population happiness index;
Shanghai focuses on the system innovation and China Shanghai Free Trade Test Zone is under construction step by step;
China's economic operation remains to be stable to be good with steady progress of structural adjustment and improving quality of the transformation and upgrading;
The world economic recovery situation is unstable with deep adjustment of the economic structure and the international competition is more and more fierce.

2015/2016秋冬海派时尚流行趋势

社会背景 SOCIAL BACKGROUNDS

科技 | TECHNOLOGY

比特币、快的打车、极路由、知乎社交

旧去新来，创新公司和创新产品层出不穷；

传统科技巨头面临诸多挑战，"转型"成为流行语；

扑朔迷离的技术现象和创意时尚的完美结合，带来意外的惊喜；

正在路上的中小创业者，代表了科技和时尚行业的未来。

Bitcoin, taxi booking apps, pole routes and quaro social networking …
The innovative companies and innovative products emerge endlessly from old to new;
The traditional technology giants are faced with many challenges, and "transformation" has become a buzzword;
The perfect combination of the most complicated and confusing technological phenomenon and creative fashion brings the unexpected surprises;
The small and medium entrepreneurs are on the way and they represent the future of technology and fashion industries.

2015/2016 AUTUMN/WINTER STYLE SHANGHAI FASHION TREND

社会背景 / SOCIAL BACKGROUNDS

环境 | ENVIRONMENT

北涝南旱、高温瞠目、十面霾伏……
气候变化和环境恶化的匆匆脚步声，叩响每个人的心房；
向空气、阳光、河流道歉，向环境治理痼疾宣战，低碳发展路线图科学制定；
上海着力推动长三角区域大气污染联防联控，百万家庭低碳行……
公民之权、企业之事、政府之责，保护蓝天净土，家园依旧美好。

Droughts in the north, floods in the south, startling high temperature, hazy weather...
The rush footsteps of the climate change and environmental deterioration knock on everyone's atria;
Make an apology to the air, sunshine and river and fight with the environmental governance problems
with the scientific development of low-carbon roadmap;
Shanghai focuses on promoting the joint prevention and control of air pollution in the Yangtze River Delta
with millions of families of low-carbon behaviors...
It is the citizen right, corporate matters and government duties to protect the blue sky and pure land and our home is still beautiful.

2015/2016秋冬海派时尚流行趋势

社会背景 SOCIAL BACKGROUNDS

生活 | LIFESTYLE

狙击房价快速上涨，时尚休闲消费占比提高；

教育改革试行、为老服务水平提升，百姓幼有所教、老有所养、病有所医；

衣食住行一网扫尽，体验指尖上的智慧生活；

当时尚遇见科技，生活不仅是生存。

Sniper the rapidly rising house prices and the proportion of fashion leisure consumption is increased;

Enhance the level of services for the people to provide education for the kids, support for the old and hospitalization for the patients with trial implementation of education reform;

The basic necessities are all included and experience the wisdom of the life on the fingertips;

When the fashion meets the technology, life is not just for the survival.

2015/2016 AUTUMN/WINTER STYLE SHANGHAI FASHION TREND

社会背景 / SOCIAL BACKGROUNDS

艺术 | ART

国务院推进文化创意和设计服务相关产业融合发展；

上海自贸区开放文化服务新举措，预示文化新业态；

艺术开始大众化消费，观展看剧成为日常生活一部分，文化更惠民；

嫁接资本增强软实力，中国跻身全球最重要艺术市场，文化正大踏步走出去；

上海国际文化大都市格局凸显，海派文化焕发新活力。

The State Council promotes the integrative development related to the cultural and creative designing industries;
Shanghai FTA opens up a new measure for the cultural services, indicating a new form of culture; The art tends to be a mass consumption.
As the exhibition concept and drama becomes one part of daily life, the culture is more and more close to the people;
The grafting capital enhances the soft power.
As China ranks among the world's most important art market, the culture is making great strides;
The pattern of Shanghai international cultural metropolis is highlighted and the Shanghai Style Culture is displayed with a new vitality.

2015/2016秋冬海派时尚流行趋势

趋势演绎 TREND INFERENCE

革新·生机 RENOVATION·VITALITY

塑 creat
[海派经典风格]
SHANGHAI CLASSIC STYLE

续 sustain
[海派自然风格]
SHANGHAI NATURAL STYLE

经典时尚注入时代体温
传统元素重新多元演绎
东西碰撞再塑时代经典
摩登现代焕发海派风韵

时尚设计重回健康生活
温和力量创造和谐设计
传统人文焕发时代光彩
乐活精神渗透都市生活

Classic fashion integrates with the temperature of era
Traditional elements present the multiple reinterpretation
Eastern and Western collision reshapes the classical era
Modern style glows the charm of Shanghai Style

Fashion design returns to the healthy life
Moderate forces create the harmonious design
Traditional humanism glows the glory of era
LOHAS spirit penetrates into the urban life

2015/2016 AUTUMN/WINTER STYLE SHANGHAI FASHION TREND

趋势演绎
Trend Inference
革新·生机 RENOVATION·VITALITY

趣 fun
[海派都市风格]
SHANGHAI URBAN STYLE

变 variety
[海派未来风格]
SHANGHAI FUTURISTIC STYLE

东方巴黎散发年轻活力
多元融合带来创意革命
艺术与设计更贴近生活
国际环境重塑传统乐趣

技术与艺术的完美结合
智能体验引发先锋探索
功能需求带来便捷时尚
优化再造引领科技革新

Oriental Paris presents the youthful vigor

Diverse fusion brings the creative revolution

Artistic design is closer to life

International background reshapes the traditional fun

Technology and art are perfectly combined

Intelligence leads to the pioneering exploration

Functional demand brings the convenient fashion

Optimized recycling leads to the technology

2015/2016秋冬海派时尚流行趋势

趋势演绎 TREND INFERENCE

革新·生机 RENOVATION·VITALITY

塑
creat
[海派经典风格]
SHANGHAI CLASSIC STYLE

2015/2016秋冬海派时尚流行趋势主题
2015/2016 AUTUMN/WINTER STYLE SHANGHAI FASHION TREND THEMES

革新·生机 RENOVATION·VITALITY

2015/2016秋冬海派时尚流行趋势

主题说明 THEME DESCRIPTION

革新·生机 RENOVATION·VITALITY

塑 creat [海派经典风格 SHANGHAI CLASSIC STYLE]

塑
中文拼音：sù
英文：creat，mould，shape
组词：塑造，塑形，再塑，雕塑
趋势分类：海派经典风格
趋势简述：对海派文化的传承和重塑

　　在2015/2016秋冬"革新·生机"的流行趋势主题下，海派经典风格所演绎的是对海派传统文化的重新设计与革新，从而呈现出"新海派时尚"的勃勃生机。

　　海派经典风格由传承积淀的典雅海派故事和国际时尚流行要素共同构成。所谓经典，就是经久不衰，世代留芳。昨日的时尚精华成为今日的经典，而今日的流行时尚是对昔日经典的再次塑造。海派故事来自于外滩、石库门、旗袍、百乐门、月份牌美女等人尽皆知的上海印象，而国际时尚流行来自于巴黎、纽约、伦敦最新的时尚潮流，两者的结合塑造出既是本土的，又是世界的设计新作，这种融汇东西、海纳百川的气质，亦是海派时尚的精髓所在。

　　从创意策划大师邵隆图设计的石库门老酒"海上繁华"系列，与海派画家李守白跨界合作，到隆图先生在上海家化"双妹"品牌中所塑造出的新海派时尚女性形象；从服装设计师武学凯的作品"华彩"，到老凤祥品牌兼具典雅和创新的海派时尚首饰，无不渗透着对于中国传统文化、上海经典元素的尊重和传承，充分展现了在当今的时代背景下的时尚气质。

2015/2016 AUTUMN/WINTER STYLE SHANGHAI FASHION TREND

主题说明
THEME DESCRIPTION

革新·生机 RENOVATION·VITALITY

塑 creat
[海派经典风格]
SHANGHAI CLASSIC STYLE

Creat
Chinese Pinyin: sù
English: create, mould, shape
Trend category: Classic Shanghai Style
Trend description: inheritance and remodeling to Shanghai Style Culture

  Under the theme of the "Renovation · Vitality" fashion trend in the 2015/2016 autumn and winter, the Classic Shanghai Style demonstrates the reinterpreted design and innovation to the traditional Shanghai Style Culture so as to present the vitality of the "New Shanghai Style Fashion".

  The Classic Shanghai Style is composed of the elegant Shanghai style stories with an accumulation of inheritance and the international fashion elements. The so-called classic is lasting and to achieve the immortality. The fashion essence yesterday has become the classic today and today's fashion is once again the shaping for the old classic. The Shanghai style stories come from the Bund, Shikumen, Cheongsam, Paramount, Calenda Beauty and other well-known Shanghai impressions and the international fashion trends come from the latest fashion trends of Paris, New York and London, whose combination has formed the pieces of both local and international design works. This kind of temperament integrating the east and west is tolerant to diversity and it is also the essence of the Shanghai Style Fashion.

  From the cross-border cooperation between the Shanghai style painter Li Shoubai and the creative planning master Shao Longtu who designed the "Prosperous Sea" series of Shikumen old wine; to the female images of the new Shanghai Style Fashion created by Mr. Shaotu in the brand of Two Girls of Shanghai Jahwa; from the works of "Hopewell" on the opening of the Shanghai Fashion Week made by the costume designers, Wu Xuekai and Wu Xuewei, to the elegant and innovative Shanghai Style Fashion jewelry of Old Phoenix brand; all of them are permeated with the respect for the tradition and cultural heritage of China and the classic Shanghai elements with a full demonstration to the fashion style in the background of today's era.

2015/2016秋冬海派时尚流行趋势

趋势演绎 TREND INFERENCE

革新·生机 RENOVATION·VITALITY

塑 creat
[海派经典风格] SHANGHAI CLASSIC STYLE

本主题的色彩充满传统韵味。内在的品质感与优雅感是这一系列色彩表现的主要方向。经典的奶酪色在奢华的祖母绿、深绛紫等色彩的点缀下变得丰盈时尚。

The color in this theme is full of traditional charm. The inherent sense of quality and elegance is the main direction of color performance in this series. The classic cream color becomes fashionable and rich in the embellishment of luxury emerald, deep plum and other dark colors.

2015/2016 AUTUMN/WINTER STYLE SHANGHAI FASHION TREND

主题色彩
COLOR THEME

珍珠白 SH1200022	砖灰色 SH0401902
浅奶酪 SH0700516	赤金色 SH0200513
奶油驼 SH0700407	木兰黄 SH0101264
墨水蓝 SH0503622	深酒红 SH0303671
苍灰蓝 SH1101108	深鸦青 SH1101064
黛蓝色 SH0503408	月白蓝 SH1101105
祖母绿 SH0902958	海棠红 SH0201808
浅绾色 SH0303096	深绛紫 SH0402746
紫檀棕 SH0701453	薰衣草 SH0402289
胭脂红 SH0303636	青紫蓝 SH0402252

2015/2016秋冬海派时尚流行趋势

塑creat
[海派经典风格]
SHANGHAI CLASSIC STYLE

邵隆图
Shao Longtu

九木传盛品牌推进机构董事长
上海市品牌建设工作联席会议专家委员会委员
上海国际时尚联合会副会长
上海创意设计中心首席策划大师
上海创意产业中心专家委员会副主任
东华大学服装·艺术设计学院兼职教授

中西合璧的海派繁华

石库门是20世纪二三十年代的上海典型民居。经几代人的打造，石库门已经成为中西合璧、海纳百川上海城市精神的象征。石库门上海老酒的设计不失时机地把中西合璧的海派城市精神，融入产品系列设计，成为始终不变的石库门品牌风格。

2013年1月，石库门品牌推出"海上繁华"系列，与海派画家李守白跨界合作，推出限量礼品酒，分别用六幅画装饰，提升了石库门品牌的艺术价值。

"海上繁华"系列礼盒设计，象征红底砖墙的盒面上，六幅画组成一件彩色的旗袍，海派文化的时尚感更为强烈。

换行头　　寻搭子　　蟹红了

绣花头　　讨媳妇　　着棋子

"海上繁华"系列礼盒设计

2015/2016 AUTUMN/WINTER STYLE SHANGHAI FASHION TREND

[海派经典风格] SHANGHAI CLASSIC STYLE

STYLE SHANGHAI PROSPEROUS WITH ELEGANT SMART

Shikumen was the typical residence in the 1920s and 1930s of Shanghai. After the construction of several generations, it has become symbol of Shanghai's spirit boasting a combination of Chinese and Western elements with a tolerance to diversity. The design for the old wine of Shanghai Shikumen seized the opportunity to incorporate it into the product information design, which made it to always be the same brand style of Shikumen.

In January 2013, the Shikumen brand launched the series of "Prosperous Sea" and had a cross-border cooperation with the Shanghai style painter, Li Shoubai, to offer a limited gifts of liquor, which was separately decorated with six paintings to enhance the artistic value of Shikumen brand.

The gift box design of the "Prosperous Sea" series was six paintings to form a color cheongsam on the box surface symbolized with the red brick wall in the background, making the fashion of the Shanghai Style Culture more intense.

"双妹"品牌 重逢上海

又嗲又㜺的东情西韵

2008年，上海家化复活"双妹"老牌，华丽转身取名"Shanghai VIVE"。传承"双妹"老月份牌风格，创新出"双妹"彼时名媛，转身留芳，幻成万般风采的新品牌形象。凸显海派名媛又嗲又㜺（jia）的名媛气质。

THE DIAISTIC AND CAPABLE EASTERN SENTIMENT AND WESTERN CHARM

Shanghai Jahwa revived the old brand of "Two Girls" and it was renamed to be "Shanghai VIVE" with a gorgeous turning in 2008. The newly created "Shanghai VIVE" adheres to the style of the "Two Girls" old brand and it is now full of the elegant quality of the famous ladies with various brand images, which highlights the diaistic and capable elegant quality of famous ladies in Shanghai.

2015/2016秋冬海派时尚流行趋势

塑 creat
[海派经典风格] SHANGHAI CLASSIC STYLE

武学凯
Wu Xuekai

中国服装设计师协会常务理事、主任委员
上海服装设计协会副秘书长
上海市高级服装设计师
上海工业设计协会常务理事
上海青年文联理事
上海市青年高端创意人才协会理事
亚洲时尚联合会理事
天津工业大学艺术设计学院客座教授
上海赢国服饰有限公司创意总监
上海标顶服饰有限公司创意总监

时装设计中的东方哲学

设计的过程就是发现的过程，是一个利益和功能不断叠加的过程。是科学、艺术、情感的物化，这就是今天设计师的工作。我们通过学习、体验、感受、感知、交流、总结和提炼，不断寻找创新的机会点、利益点。

希望人们在多元的时代里，把持着各自的信仰、文化和自我的独立性。"华彩"是矛盾多样性的统一，是不同事物间的自然平衡。系列作品"华彩"以中国传统元素为灵感，运用当代时尚话语，诠释对人与人、人与自然合而不同，分而共生的智慧。

华彩系列服装设计

2015/2016 AUTUMN/WINTER STYLE SHANGHAI FASHION TREND

ORIENTAL PHILOSOPHY IN FASHION DESIGN

The design is the process of discovery and it is a superimposed process of the benefits and functions. It is the materialization of the science, art and emotions. This is the designer's works today. We are constantly looking for opportunities and benefits for innovation by learning, experiencing, feeling, perception, communication, summarizing and refining.

It is expected that people can adhere to their own faith and culture, while maintaining the self-independence and being inseparable with others in the pluralistic era. The "Huacai" is the unity in the contradictory diversity and the natural balance between different things. The "Huacai" series of works are inspired by the Chinese traditional elements. to interpret the difference between the human beings and the difference between the nature and the human, as well as the wisdom in the separate mutualistic symbiosis by using the modern stylish words.

华彩系列服装设计

2015/2016秋冬海派时尚流行趋势

[海派经典风格]
SHANGHAI CLASSIC STYLE

老凤祥名师设计中心
SHANGHAI LAOFENG CO.,LTD

上海市原创设计大师工作室

以年轻之心传承海派首饰

　　珠宝设计是一件纳须弥于芥子的艺术，宏大的人类主题、深远的人生领悟、对美好生活的期许、浓浓的感情都可被容纳于这颈间、指上、耳畔缠绕的小物之中。你于昂贵的宝石、巧夺天工的工艺之中看到的，正是大千世界的投影。海派首饰意味着海纳百川、包罗万象，容纳各种文化的精髓。老凤祥165年来一直秉承着这样的精神，从解放前为宋氏三姐妹、哈同夫人等一批社会名流定制设计的首饰，到如今装点时尚女性的配饰，始终不变的是一种细腻的美、精致的美、巧妙构思的美，这种美与上海人的生活情趣、上海女人的优雅韵致一脉相承。"老凤祥名师设计中心"是于2004年由上海市经济委员会首批授牌成立的"上海市原创设计大师工作室"之一，是百年老字号——"老凤祥银楼"旗下的一支精英设计团队。在这个团队里，各种文化、各种思想在激情碰撞，创意火花四溅，给予了老凤祥品牌新鲜而富有活力的血液。

花开富贵

鸾凤祥和

富贵幽兰

2015/2016 AUTUMN/WINTER STYLE SHANGHAI FASHION TREND

HERITAGE OF SHANGHAI JEWELRY STYLE WITH YOUNG HEART

龙图腾檀木银饰

竹翠春晓

炫舞飞天

[凤系列] 龙凤吉祥

Jewelry design is an art of integration and harmony. All the themes of great mankind, profound comprehension of life, hope for better life, deep feeling can be integrated in the small objects on the collar, fingers and ears. You may appreciate the wider world from the expensive gems and intricate process. Shanghai style jewelry refers to embracing the diversity and containing the essence of various cultures. Laofengxiang has been adhering to this spirit in its history of 165 years, from the custom jewelries for the Soong Sisters, Mrs. Hardoon and a number of celebrities before the liberation in 1949, to the decorative accessories for fashionable women, presenting the same delicate, exquisite, ingenious beauty at all times. This beauty is in harmony with the delightful life of Shanghai people and the elegant charm of Shanghai women. Founded in 2004, Laofengxiang Designer Center is one of the first "Shanghai Original Design Master Studios" approved by Shanghai Municipal Economical Commission, and is an elite design team of Laofengxiang Jewelry, a century-old brand. In this team, the passionate collision of various cultures and ideas produces the great creativity, giving the fresh and vibrant energy to the brand of Laofengxiang.

2015/2016秋冬海派时尚流行趋势

塑 creat
[海派经典风格]
SHANGHAI CLASSIC STYLE

HATTERS' HUB
帽仕汇

知名帽饰品牌

精致帽饰中的中国元素新演绎

热爱帽饰的人都知道帽仕汇,热爱帽仕汇的人都狂热地追随着帽饰的动向。一个在全球享誉盛名的帽饰设计公司在2015/2016的秋冬为我们带来了精致而不失创意的新设计。细巧的刺绣,经典的蝴蝶图案,轻巧的羽毛等都是流行方向盘上的小按钮。

秋冬的寒冷是带不走女神的魅力的。游走于社交与家庭的女性一定不会缺少一款细巧钩编,精致装饰的帽饰。蜿蜒的线条缠绕在女性的头顶,深沉的暗红诉说这刚强外表下的柔软内在。网纱、波点和暗夜羽毛平添了一抹趣味的妖娆。

JEFF SUN 帽仕汇

2015/2016 AUTUMN/WINTER STYLE SHANGHAI FASHION TREND

[海派经典风格] SHANGHAI CLASSIC STYLE

NEW CHINESE ELEMENTS IN ELEGANT HAT DESIGN

People who love hats all know about the HATTERS' HUB and people who love HATTERS' HUB are feverishly in pursuant of the hat trends. One company that is globally renowned for the hat designing had brought us delicate new design without any loss of creativity in the 2015/2016 autumn and winter. The small embroidery, classic butterfly pattern and light feathers and so on are all the small buttons on the steering wheel of the fashion.

The charming of the goddess will not be taken away by the cold winter. Women who are busy with the society and family will never lack a delicate decorative hat with exquisite crocheting. The meandering lines are wrapped around the woman's head and the deep dark red tells the soft heart under the strong appearance. The gauze, polka dots and dark feathers has added a touch of interesting enchanting.

首届中国帽仕汇杯帽饰设计大赛作品

首届中国帽仕汇杯帽饰设计大赛作品

2015/2016秋冬海派时尚流行趋势

[海派经典风格]
SHANGHAI CLASSIC STYLE

红谷皮具有限公司

传统手工的传承与发扬

  优美而富于联想的名字，它诞生在美丽的丽江，汲取丽江清澈的丽水，呼吸丽江沁人心脾的空气，采摘丽江本土淳朴气息，最终诞生出了诗意而柔美的女性箱包领导品牌。精于工艺，巧于设计让红谷在新一代独立女性的心中留下了难以磨灭的影响。

  在当今传统手工艺匮乏的年代，我们在红谷可以看到其优美、精巧的设计。镂空刻画，精雕细琢，仿佛看到了女性的细声细语和婀娜身姿。东方女性的柔美就在这一个个精致的工艺下渐渐显现出来了。

产品工艺

2015/2016 AUTUMN/WINTER STYLE SHANGHAI FASHION TREND

[海派经典风格]
SHANGHAI CLASSIC STYLE

creat塑

THE INHERITANCE AND PROMOTION OF THE TRADITIONAL HANDWORK

Being a beautiful name full of imagination, it was born in the beautiful Lijiang. It draws from the clear Lijiang water, breaths the refreshing Lijiang air, picks up the simplicity of the local Lijiang, and eventually the leading brand of poetic and gentle women's bags & cases is established. It is skilled in the crafts with a clever design that makes HONGU brand has left an indelible impression on the new generation of independent women.

In today's era of scarcity of traditional arts and crafts, we can see its beautiful and sophisticated design in the HONGU. It seems that you have heard the gentle women's voice and seen the gentle figure with the pierced engraving and delicate craft with uncompromising attention. The gentle beauty of the oriental women is gradually emerged with these exquisite craftsmanships.

"青瓷"系列

2015/2016秋冬海派时尚流行趋势

塑 creat
[海派经典风格] SHANGHAI CLASSIC STYLE

上海时湾艺术设计有限公司

传统形象的时尚演绎

"青红"以老上海的社会生活为题材,挖掘民间的传统记忆,以图形模块为基础进行文化创意产品的衍生开发,演绎老上海人物群的生动特征和怀旧风情。衍生形式有:文化读物、T恤、DIY玩具、文具、动画、手游、公共艺术等,在形式上具有持续衍生的能力。"中品"以古典的江南文化为题材,以抽象提炼出的江南元素重新演绎具有时尚色彩的现代图形及文创衍生品。专注于挖掘江南文化的特征,结合国人回归的自我意识和日常的功能需求进行开发,注重模块化研究和整体开发上的市场化预设。

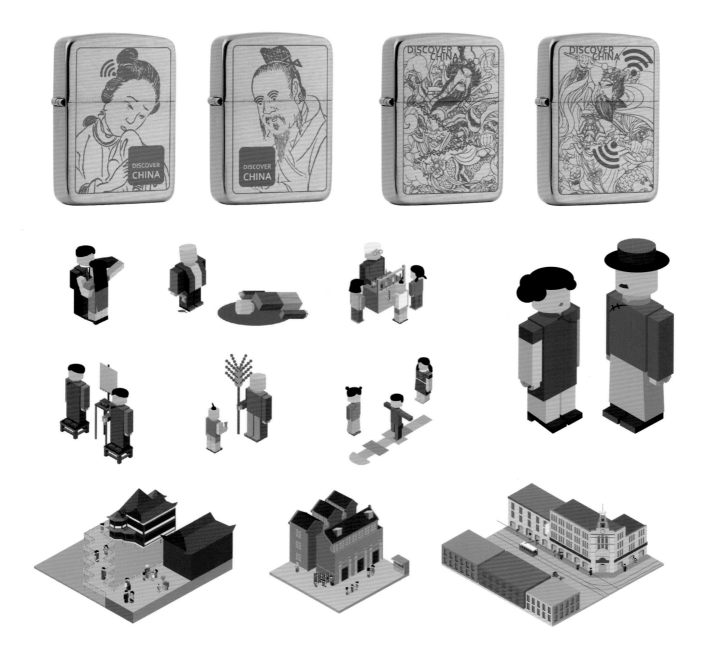

FASHION INTERPRETATION OF TRADITIONAL IMAGE

The "red green" takes the social life in the old Shanghai as the theme to tap the traditional folk memory so as to have the derivative development on the cultural and creative products and interpret the vivid features and nostalgia of the people in the old Shanghai. Derivative forms: cultural books, T-shirts, DIY toys, stationery, animation, mobile games, public art, etc. It has the sustainable generative capacity in the form. The "Zhongpin" is about the classic Jiangnan culture to reinterpret the modern graphics and the cultural and creative derivatives with the stylish colors by way of the abstract Jiangnan elements. It focuses on tapping Jiangnan cultural features to have the development by way of combining with the Chinese people's self-awareness and daily functional requirements. It focuses on the modularized research and the market – oriented presetting of the overall development.

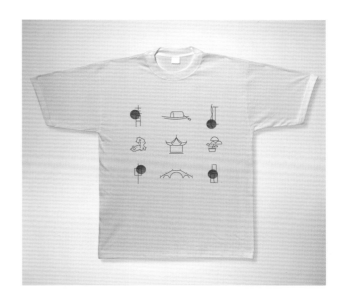

2015/2016秋冬海派时尚流行趋势

[海派经典风格]
SHANGHAI CLASSIC STYLE

海派文化符号的现代应用

在当今时代，传统的样式在设计师手中可能被重新解读。在这种大胆解读中，它们的形象虽还带有明显的文化特征，但其内容或意义已经有了全新演变。以历史典故"武松打虎""狮子头""庄子""西游记"等做为源泉，这些画面已经成为时尚的现代的图形。

"喜立方"文创产品系列将中国传统的双喜图样做为基本元素，通过用不同材料和结构进行演绎，衍生出了多元化的系列产品。如结合座椅、书立、书签、名片盒等使用功能，使普通的产品巧妙融合文化符号后具备了独特魅力。

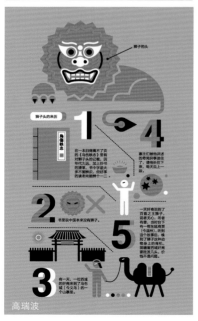

高瑞波

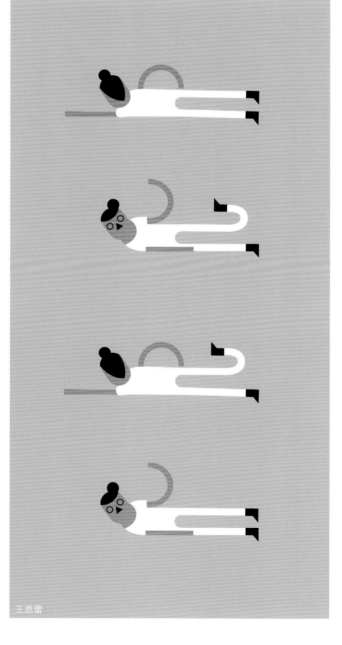

王思蕾

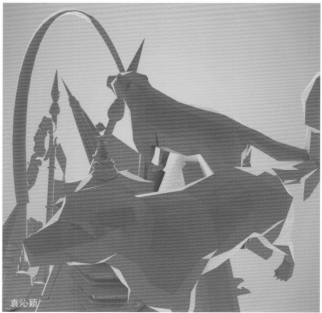

袁沁颖

2015/2016 AUTUMN/WINTER STYLE SHANGHAI FASHION TREND

上海洋滔品牌策划有限公司

THE MODERN APPLICATIONS OF SHANGHAI STYLE CULTURE SYMBOL

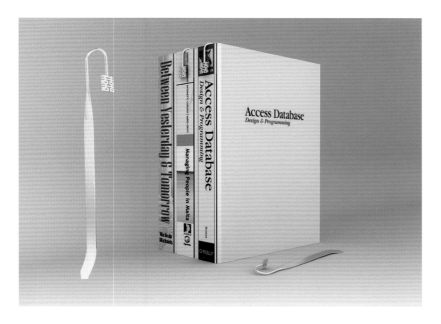

In today's era, the traditional style may be re-interpreted by the designers. Though their images are still with the clear cultural features, the content or meaning has been totally evolved in such a bold interpretation. Using the historical allusions "Wu Song fought the tiger", "lion head" and "Zhuangzi", "Journey to the West" and so on as the source, these images has become the modern and stylish graphics.

The "cubic happy" cultural and creative products series use the traditional Chinese double happiness pattern as the basic elements to have the interpretation with different materials and sharps, which has derived into the diversified products, such as the ordinary products combined with the chairs, book stand, bookmarks and card box and other using functions to be with the unique charming after the clever integration with the cultural symbols.

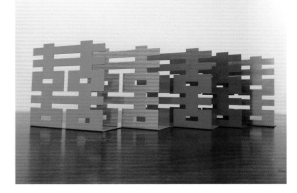

塑 creat
[海派经典风格]
SHANGHAI CLASSIC STYLE

厚重中凝练时尚之美

城市的辉煌经历渗透在温润、沉着的系列面料中，吞吐万汇、处杂不凡。骄傲的兼具经典、精致的追赶时髦，展现了时尚性、功能性的特征。

雕塑造型的装饰效果，通过梭织仿针织的精纺条呢演绎英伦冷峻，与法兰西浪漫格纹穿越、巴洛克雄奇、洛可可神经质混搭一本正经的图案，精心策划、自由组合成随意漫不经心，具有毛糙表面、却密实的呢子、针织物中。各种功能性纤维大行其道，仿棉纤维、仿毛纤维、保暖纤维令人舒心。

沈沉
Shen Chen　水舞深造

东华大学服装·艺术设计学院
纺织品艺术设计专业教研室主任
上海水舞深造文化传播有限公司创意总监
国家纺织产品开发基地·特聘评估专家

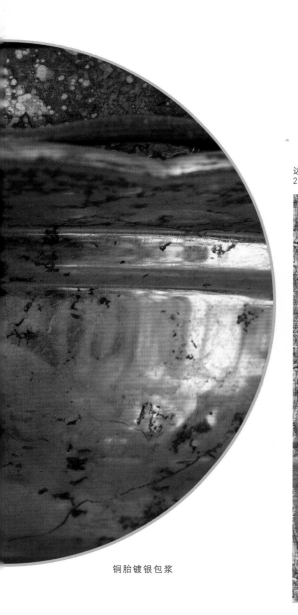

铜胎镀银包浆

达利丝绸　沈沉《520》
250g/㎡，黏胶74%，桑蚕丝26%，梭织印花

ZOOUZA　沈沉《巴赫的管风琴声》
615g/㎡，羊毛70%，锦纶30%，针织提花

水舞深造　沈沉《到访》
482g/㎡，桑蚕丝71%，羊毛23%，羊绒6%，色织

[海派经典风格] SHANGHAI CLASSIC STYLE

AESTHETIC OF STYLE IN STRENTH

The urban brilliant experience is penetrated into the tender and cool fabrics, miscellaneous and extraordinary. The proud pursuit of classic and refined style is a show of fashionable and functional characteristics.

With sculpture-like decorative effect, the refined pongee by knitting and stitching is mixed with the solemn British style, romantic French plaid, magnificent Baroque and Rococo neurotic patterns, to freely form the casual styling of sense by woolen and knitted fabrics with rough surface. Various functional fibers are popular to give the soothing of cotton fibers, wool fibers and warm fibers.

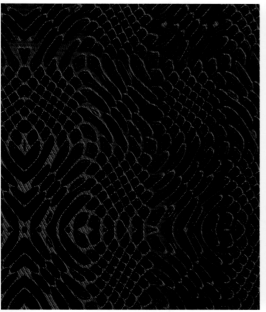

水舞深造　沈沉、狄晨光《仁义道德》
578g/㎡，黏胶46%，羊毛40%，腈纶20%，发泡涂层

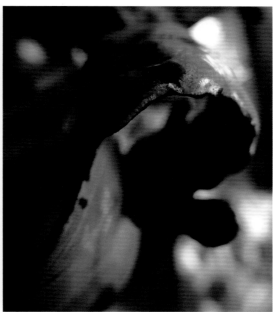

历久弥新的搪瓷

水舞深造　沈沉、刘静妍《曲意》
552g/㎡，氨纶31%，黏胶46%，羊毛7%，腈纶16%，转印花

水舞深造　沈沉、张安琪《经典重述》
460g/㎡，变性腈纶29%，转印花

鼎天时尚　《肃蓝巴洛克》
317g/㎡，氨纶31%，黏胶46%，羊毛7%，腈纶16%，梭织提花

2015/2016秋冬海派时尚流行趋势

塑 creat
[海派经典风格] SHANGHAI CLASSIC STYLE

倪明
Ni Ming

东华大学服装·艺术设计学院
纺织品艺术设计专业教师

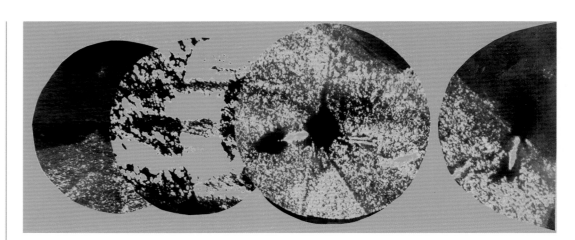

海上风景之一

塑造有意味的形式

  有意味的形式体现在形式构成要素点线面和色彩用某种特殊方式组成的某种形式或形式间的关系，从而激起人们的审美感情，"有意味的形式"被认为是一切视觉艺术的共同性质。在塑造一个场景、物体时，如何塑造有意味的形式是要在整个创作过程中始终密切关注的问题。不仅要考虑轮廓结构、空间、光影等形式要素，同时也要注重表现视觉感和触觉感，因为这些方面能引发人的情感共鸣。在构图上似乎非常自由，但形式之间的关系是在冲突和调和中妥协的。除了对立形态之间的互渗互容，从触觉角度来模糊锐利的边角和清晰的质感也是处理形式关系的有效方法。

海上风景之二

海上风景之三

2015/2016 AUTUMN/WINTER STYLE SHANGHAI FASHION TREND

SHAPE THE MEANINGFUL FORM

The meaningful form which is considered to be the common nature of all the visual arts is reflected to the constituent elements of point, line and plane and a certain kind of formation formed by the some special methods with the color or the relationships between the formations so as to arouse people's aesthetic feelings. When I am shaping a scene or an object, I always pay close attention to how to shape a meaningful form in the whole creative process. Not only the elements of outline structure, space and light and shadow, etc. should be considered, it also needs to focus on the performance of the visual sense and tactile sense, because these aspects can arouse people's emotional resonance. It seems to be very free on the composition of a picture, but the relationships between the forms are compromised in the conflicts and reconciliation. In addition to the mutual capacitance between the opposite forms, it is also an effective way to deal with the formation relationships from the perspective of haptic to blur the sharp edges and clear texture.

教堂印象之二

教堂印象之一

染色面料

2015/2016秋冬海派时尚流行趋势

塑 creat
[海派经典风格]
SHANGHAI CLASSIC STYLE

俞英
Yu Ying

东华大学服装·艺术设计学院
产品设计系副主任、副教授

现代与传统相结合的配饰设计
A COMBINATION OF MODERN AND TRADITIONAL ACCESSORIES DESIGN

灵感来自中国传统服饰旗袍，旗袍在上海曾一度非常流行，歌舞升平时、优雅读书时或是烟气渺渺晨光中，旗袍展现着中国女子的曼妙东方气质。优雅的高开衩结构和盘扣作为主要的设计元素，以仙鹤牡丹这个民族气质浓郁的盘滚面料搭配现代与神秘皆具的黑色皮料塑造现代与传统相互融合的海派韵味。

It is inspired by traditional Chinese cheongsam that was used to be very popular in Shanghai. The cheongsam demonstrates the graceful oriental temperament of Chinese woman when she is singing and dancing; when she is reading elegantly; or when she is in the dim morning light. The elegant structure with high slits and plate buttons are the main design elements. The rolling materials of crane and peony with strong temperament of this nation are used to match with the Modern and mysterious black leather to shape the Shanghai style charm integrated with modernity and tradition.

汝海洋 鞋履设计作品
民族刺绣图案的立体构成
中国传统绣花布鞋
老街巷落的咖啡厅
传统竹质折扇

2015/2016 AUTUMN/WINTER STYLE SHANGHAI FASHION TREND

[海派经典风格]
SHANGHAI CLASSIC STYLE

创

creat

徐遥 箱包设计作品

包在款式面料上，特别采用了中国传统的"双凤朝牡丹图"旗袍绸缎面料以及纯天然的环保麻质面料，并结合当下所流行的潮牌文化中的个性、叛逆、大胆等设计风格特点进行搭配设计。在结构上面，采用了复合式的多层次设计，这样可以使使用者在装放物品时对其进行有效归类，分类装放，这样就可大大减少使用者因所有物品都混乱堆放在一起，找时却找不到这样的烦恼了。在双肩包的肩带设计上，还特别采用了更为符合人体上身曲线的设计，这样的曲线更贴合人体上半身身的表面，使包包背在身上更为轻松舒适。

The traditional Chinese cheongsam silk fabrics with "picture of double phoenix towards the peony" and natural green linen fabric are specially used on the style and materials of the bags and the personal, rebellious, bold and other design features in all the current popular culture brands are combined to match with the design. The composite multi-level design is used in the structure so that the users may have an effective classification on the goods. The classified storage can greatly reduce user's confusion because of the disordered storage of all the items and there will be no trouble for the user that the items cannot be found immediately. The more ergonomic design is specially used on the strap design of the shoulders bag so that curve is more close to the upper surface of the body to make the bags more relaxed and comfortable with the bags on the shoulder.

2015/2016秋冬海派时尚流行趋势

趋势演绎
TREND INFERENCE
革新·生机 RENOVATION·VITALITY

续
sustain
[海派自然风格]
SHANGHAI NATURAL STYLE

2015/2016 AUTUMN/WINTER STYLE SHANGHAI FASHION TREND

趋势演绎
TREND INFERENCE

2015/2016秋冬海派时尚流行趋势主题
2015/2016 AUTUMN/WINTER STYLE SHANGHAI FASHION TREND THEMES

革新·生机 RENOVATION·VITALITY

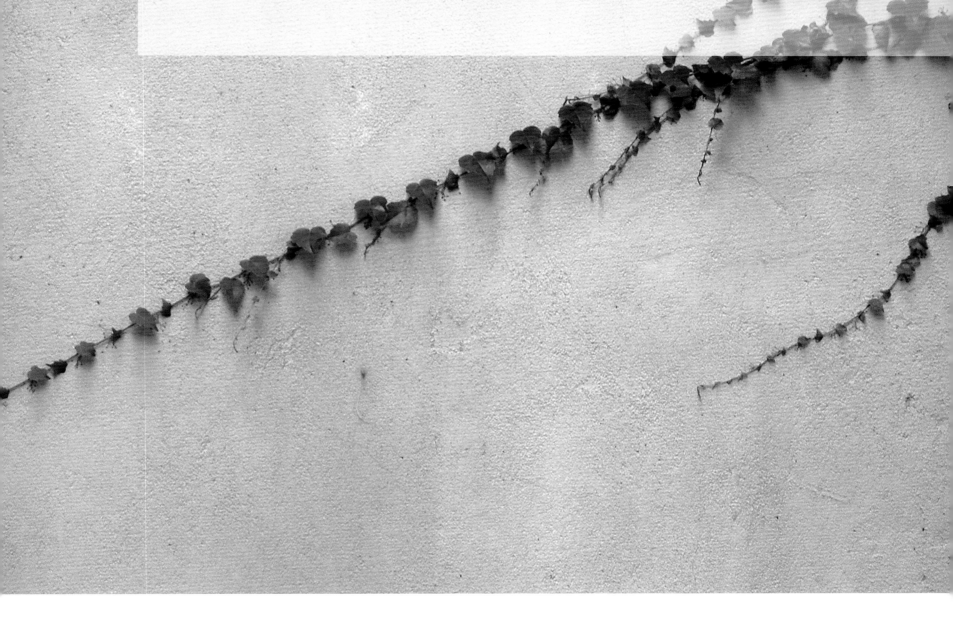

2015/2016秋冬海派时尚流行趋势

主题说明 THEME DESCRIPTION

革新·生机 RENOVATION·VITALITY

续 sustain
[海派自然风格 SHANGHAI NATURAL STYLE]

续
中文拼音：xù
英文：sustain，continue，extend
组词：继续，持续，连续
趋势分类：海派自然风格
趋势简述：自然和谐的可持续设计

  在2015/2016秋冬"革新·生机"的流行趋势主题下，海派自然风格所演绎的是尊重自然的可持续设计的革新，通过人与自然的平衡发展焕发设计的生机。

  海派自然风格体现了江南水文化，回归水乡、弄堂、市井的慢生活，重拾儿时的朴素记忆、温暖的家庭手工、原始的自然生态。在各种自然元素中,持续流动的"水"多变而永恒，是体现江南文化的最好元素，东方水墨成为本季的主要设计元素。

  设计师潘剑锋认为，繁忙的生活方式使人与人之间的距离日益疏远，人们不得不面对污染、压力和许多其他问题，设计需要"真生活"。本土品牌"弄影"创作灵感来源于江南水乡，设计总监罗竞杰将极具东方韵味的水墨、扎染、褶皱等设计元素发挥得淋漓尽致。画家吴晨荣以传统的水墨为理念，结合以国际化的当代抽象绘画语言与表现方式，充分表达了中国书法的精神与东方美学的审美特征。建筑机构FTA创始人施道红认为中国的未来是一种关注品质，关注可持续，关注健康，并能够引领我们内心精神层面的一种追求。海派自然风格中倡导的可持续设计蕴含着东方哲学，也是全球设计师之共同追求。

2015/2016 AUTUMN/WINTER STYLE SHANGHAI FASHION TREND

主题说明
THEME DESCRIPTION
革新·生机 RENOVATION·VITALITY

续
sustain
[海派自然风格]
SHANGHAI NATURAL STYLE

Sustain
Chinese Pinyin: xù
English: sustain, continue, extend
Group of words: continuing, sustaining, succession
Trend category: Natural Shanghai Style
Trend description: natural harmonious sustainable design

Under the theme of the "Renovation · Vitality" fashion trend in the 2015/2016 autumn and winter, the Natural Shanghai Style demonstrates the natural harmonious sustainable design through a balanced development between man and the nature to enliven the design.

The Natural Shanghai Style reflects the Jiangnan water culture, returning to the slow pace of life in the water villages, alleys, streets and marketplaces to regain the simple memories in the childhood, warm household handcrafts and the original natural ecology. In a variety of natural elements, the continuous flow of "water" is changeable and eternal and it is the best element to represent the Jiangnan culture, and the oriental ink has been the major design element in this season.

The designer Pan Jianfeng believes that busy lifestyle has made people become more and more strange to each other and people have to be faced with the pollution, stress and many other issues, while the design needs a "real life." The local brand "beautiful shadow dancing" is inspired by the Jiangnan water villages. The ink, tie-dye, folding and other design elements full of the oriental charm were put into full use by the director of design, Luo Jingjie. The painter, Wu Chenrong, takes the traditional ink as the concept and combines with the contemporary international abstract painting language and expressions to give a full expression to the spirit of Chinese calligraphy and the aesthetic characteristics of the oriental aesthetics. The building institutions FTA founder, Shi Daohong, believes that China's future will focus on the quality, sustainability, health and will lead us to an inner spiritual pursuit. The sustainable design advocated by the Natural Shanghai Style implies the oriental philosophy and it is also the common pursuit of the global designers.

2015/2016秋冬海派时尚流行趋势

趋势演绎 TREND INFERENCE

革新•生机 RENOVATION•VITALITY

续 sustain
[海派自然风格] SHANGHAI NATURAL STYLE

本主题从大自然中汲取灵感，暖灰色调散发静谧感，纯正有机的绿色调、淡雅质朴的原木色彩为原本黯哑的色系带来温暖与活力。大地色系逐步往棕色偏移，成为新的流行色。

The color in this theme is inspired from nature. The warm gray tone exudes a sense of tranquility, and the pure organic green tone and elegant rustic wood color bring the warmth and vitality to the original dumb colors. The earth tone gradually shifts to brown color and becomes the new fashionable color.

2015/2016 AUTUMN/WINTER STYLE SHANGHAI FASHION TREND

主题色彩
COLOR THEME

色名	编号	色名	编号
鱼肚白	SH0303045	天空蓝	SH0500976
丁香粉	SH0401979	蟹壳黄	SH0200996
石板灰	SH1100891	秋香黄	SH0904042
卵石灰	SH0402625	橄榄灰	SH0900158
玄青色	SH0503560	海带绿	SH0903641
芝麻黑	SH1000214	草席黄	SH0101258
杏仁黄	SH0801525	橡皮红	SH0300764
苹果绿	SH0903507	原木褐	SH0701299
雪松绿	SH0902804	芦苇黄	SH0701552
苔鲜绿	SH0900169	嫣红灰	SH0801751

2015/2016秋冬海派时尚流行趋势

续 sustain
[海派自然风格] SHANGHAI NATURAL STYLE

罗竞杰
Luo Jingjie

东华大学服装·艺术设计学院
服装艺术设计系副系主任
全国十佳服装设计师
弄影品牌设计总监

记忆中的江南

弄影（neoen）致力于将东方艺术人文精神与国际化流行时尚巧妙融合，专属为35岁左右有独立个性的知识女性塑造其品味形象的原创品牌。具有糅合了大气的简约风格、含蓄的东方神韵、唯美的艺术气质、朴素的自然主义四大特征的浑然天成的品牌风格。

通过品牌的五大视觉元素"水、雾、影、画、质"来解析"江南"的精髓。分别从题材、图案、抽象纹理、色块、肌理等几方面来多层次构成主题的全方位设定。把仿旧的手工艺术，如冷染、手缝线、补丁、编结、手绣、揉皱进行创新的开发。

2015/2016 AUTUMN/WINTER STYLE SHANGHAI FASHION TREND

[海派自然风格] SHANGHAI NATURAL STYLE

sustain 续

MEMORY OF JIANGNAN

The beautiful shadow dancing is committed to the ingenious fusion of the oriental art humanistic spirit and the international fashion trend and it is exclusive for the independent and intellectual women with the age of about 35 years old to shape the original brand of their own taste and image. It is a natural brand style integrated with the four characteristics of a blend of minimalist style, subtle oriental charm, beautiful artistic temperament and simple naturalism.

The essence of "Jiangnan" is analyzed by the five visual elements of the brand, that is the "water, mist, shadow, painting and quality". The comprehensive theme setting is respectively adopted by the multi-level of subject matter, pattern, abstract texture, color, texture and other aspects. The distressed manual arts, such as cold dying, hand stitching, patches, knitting, embroidery and crumpling, etc. are used for the innovative development.

2015/2016秋冬海派时尚流行趋势

续 sustain
[海派自然风格]
SHANGHAI NATURAL STYLE

吴晨荣
Wu Chenrong

东华大学服装·艺术设计学院
美术学部副教授
著名现代水墨画艺术家

传统水墨的当代性与国际化

《心象笔记—山水》系列与《字—非字》系列是上海著名当代画家吴晨荣近期的抽象水墨作品。以传统的水墨为理念，结合以国际化的当代抽象绘画语言与表现方式，在作品中充分表达了中国书法的精神与东方美学的审美特征，在创作中充分彰显了"自由生发"与"自然而然"的哲学理念。

The "Mental Notes – Landscape" series and "Word – Non-word" series are the recent abstract ink paintings of the famous contemporary artist Wu Chenrong in Shanghai. It is combined with international contemporary abstract painting language and expression methods with the concept of the traditional pen and ink to fully express the spirit of Chinese calligraphy and the characteristics of oriental aesthetics in the works. The philosophic ideas of the "natural development and growth" and "let it be" are fully demonstrated in the works.

《字-非字》系列

《心象笔记-水墨》系列

2015/2016 AUTUMN/WINTER STYLE SHANGHAI FASHION TREND

[海派自然风格] SHANGHAI NATURAL STYLE

sustain 续

心象笔记-山水系列

THE CONTEMPORARINESS AND INTERNATIONALIZATION OF TRADITIONAL INK PAINTING

The mental imagery notes-landscape Series: the traditional aesthetic pattern, modern form language, subtle and natural expression, the mental imagery is presented between the abstract and figurative.

The word – non-word series: it breaks the shackles of the traditional glyphs. The abstract modern painting visual schemata with the writing technique and concept of the calligraphy so that the painting at ease is the natural expression of the artists' inner feelings.

vitality series: rich and gorgeous colors, vivid and active brushwork, audacious and free splash-ink. All of theses show the vital feeling.

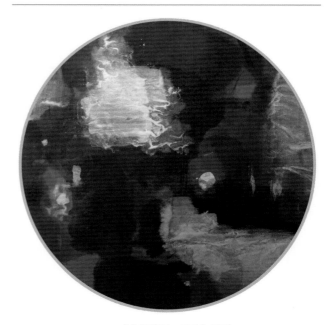

《心象笔记—天地》系列

《心象笔记—山水》系列：传统的审美格局，现代的形式语言，微妙而自然的表达，抽象与具象之间是心象的呈现。

《字—非字》系列：突破传统字形的束缚，以书法的笔法与概念，呈现出的是抽象绘画的现代视觉图式，由此随心而舞的是艺术家内心情怀的自然体现。

《生机》系列：浓郁绚丽的色彩，灵动活泼的笔法，大胆自由的泼墨，内涵的是充满生机的情怀。

《生机》系列

2015/2016秋冬海派时尚流行趋势

续 sustain
[海派自然风格]
SHANGHAI NATURAL STYLE

翟晏辛
Keeven Zhai

尚霞（上海）服饰文化传播有限公司艺术总监
知名服装设计师,造型师

欧化简约与海派风韵

  翟晏辛推崇的是简洁修身的极具女性线条感的知性优雅，展现自然与科技的融合，同时还保留了与过去古老历史的联系，"神话魅力"主题反映了返璞归真，借助了古代美人溪边作画中形象的媚态和画笔下色彩斑斓的幻象。

  古老神话中美女形象再次出现在当今社会，这正是海派文化的融合特色。面料强调线条和节奏感，铠甲的灵感设计象征着古老生活方式的胜利，并抓住了注意未来的精神，新鲜的自然色彩反映出在一个生态和谐的愿景中走向美好的户外生活。

2015/2016 AUTUMN/WINTER STYLE SHANGHAI FASHION TREND

sustain
[海派自然风格 SHANGHAI NATURAL STYLE] 续

SIMPLIFIED WESTERN STYLE SHANGHAI

Keeven Zhai advocates the simple and slim intellectual elegance that is full of the women's line sense, showing the integration with the nature and the technology while retaining the links with the ancient history. The "myth charm" theme reflects the returning to the nature by using the obsequious image in the painting of the ancient beauty beside the stream and the colorful illusion in the paintbrush.

The beautiful images in the ancient myth appear again in today's society, which is just the fusion feature of the Shanghai Style Culture. The fabrics emphasize lines and rhythm. The armor-inspired design symbolizes the victory of the lifestyle in the old times and seizes the spirit of the rayonnism future. The fresh natural color reflects a vision of ecological harmony heading to a better outdoor living.

2015/2016秋冬海派时尚流行趋势

续 sustain
[海派自然风格] SHANGHAI NATURAL STYLE

设计"真生活"

潘剑锋
Pan Jianfeng

跨文化革新设计机构—字研所SHTYPE
创作总监

我觉得时尚的东西有一部分就是来自日常的生活。今天的设计师,有可能会更多的参与到一些社会的行径中去,比如环境问题,我们需要提供一些新想法,我们需要更主动的参与其中。在茫茫人海里,环境问题、生存问题都难以避免,我在思考作为设计师在让海水变得干净一些、让大家生存得好一些中我们能做些什么。

今天我们生存在多工具的时代,掌上用户更是极其多,所以没有人对此没有责任,每个人都对环境有责任。在快速发展的现代都市中,繁忙的生活方式使人与人之间的距离日益疏远,人们不得不面对污染、压力和许多其他问题。"真生活"项目是许多来自不同文化和设计专业的人士完成的,其核心目的是探讨和发现可能出现的更好的现代中国生活。

怎么办

YOU ARE SO BEAUTIFUL

2015/2016 AUTUMN/WINTER STYLE SHANGHAI FASHION TREND

sustain 续
[海派自然风格] SHANGHAI NATURAL STYLE

"真生活"艺术家海报

"真生活"展馆

DESIGN "TRUE LIFE"

I think fashion is something that comes from a part of daily life. Today's designers are likely to be more involved in some of social behaviors, for example, the environmental issues, and we need some new ideas and some more active participation. The environmental and survival issues cannot be avoided when we are living in the open sea. I'm thinking about what we as designers can do to make the sea water become clean and make people have a better life.

Today, we live in the era with many tools. There are many palm users so everyone shall be responsible for the environment. In China's modern metropolis of rapid development, the busy lifestyle makes people become more and more strange to each other and people have to be faced with the pollution, pressure and many other problems. The "real life" project is composed of many design professionals from different cultures and its core purpose is to investigate the issue and find a better life in modern China.

日常生活写意画

2015/2016秋冬海派时尚流行趋势

续 sustain

[海派自然风格]
SHANGHAI NATURAL STYLE

施道红
Shi Daohong

FTA创始人、执行董事
同济大学建筑学硕士
米兰工学院访问学者
中欧链接设计成员
中国绿色建筑实践标兵
《时代楼盘》"金盘奖"评委
《时代建筑》理事会理事

当建筑邂逅绿色时尚

时尚不仅仅是传递一种外在的美,而是要传达出其背后的智慧。时尚需要以人的体验为中心,时尚回归健康已经成为主流。马斯洛需求金字塔正在发生戏剧性的变化,很多人满足了自我实现的需求,但却满足不了较低级别的安全需求。互联网的发展,使未来的每个人都是时尚的参与者,而不是仰望者,每个人都能够拥有时尚,时尚若不能变成一种生活方式,则其仅是一种信息而已。香奈儿大师曾说过"时尚易逝,风格永存",时尚的背后要有相当充足的价值观去引导,才具有可持续性。

FTA的思想核心是以技术和艺术思考未来。以金融街为例,FTA倡导的是一体化的设计,把建筑设计当成一座微型的城市。对于FTA而言,必须做绿色设计,这是建筑师的职责,因为建筑影响着地球40%的生态。FTA反对把建筑当成一种形式去传递美的手法。我认为中国的未来是关注品质,关注可持续,关注健康的,能够引领我们内心精神层面的追求。

花桥金融园

天津泰达慧谷服务

sustain
[SHANGHAI NATURAL STYLE] 海派自然风格 续

WHILE ARCHITECT MEETS GREEN FASHION

昆山金融街一期会所

昆山金融街

The fashion is not only to deliver a kind of external beauty but to convey the internal wisdom. The fashion needs to be people-centered experience; the fashion returning to the health has become a mainstream fashion. The Maslow pyramid dramatic changes are taking place. Many people have met the needs of self-realization, but they fail to meet the needs of lower-level security; The development of the Internet makes that everyone is the fashion participant in the future rather than the expectant person. Everyone can have the fashion and if the fashion cannot become a way of life, it is an information only; Chanel master once said, "fashion goes, style remains". It requires quite enough values for the guidance behind the fashion so that it is sustainable.

FTA's core concept is to think about the future with the technology and art. Take the Financial Street for example, FTA advocates the design of integration and the architectural design shall be considered as a miniature city. For FTA, it must do the green design. It is the responsibility of the architect, because the buildings have an impact on 40% of earth's ecology. FTA opposes to take the buildings as a form of conveying the beauty. I believe that China's future will focus on the quality and sustainable development and health and it is able to lead us to a spiritual pursuit of our heart.

昆山金融街三期

花桥国基信息城

2015/2016秋冬海派时尚流行趋势

续 sustain
[海派自然风格]
SHANGHAI NATURAL STYLE

有机形态演绎自然美感

　　海派自然可以延续此类自然风格及工艺，用色时而妖艳大气，时而朴素内敛，都呈现着源于自然的魅惑，以及设计师无限的激情与想象力。丰富多变的结构穿插，更呈现出自然梦幻的色彩。

　　随着自然资源的萎缩与生态环境的恶化，人们发现回归自然是未来人们追随的方向，采用花卉、羽毛、藤草等自然元素，结合绳编、竹编等多种工艺，可以展现出独特的造型感。每件作品中不同工艺的结合，以及不同结构的尝试运用，都体现着设计者对美好事物的探索、追求。

隋宜达
Sui Yida

帽饰设计师

枯藤傲骨系列

枯藤傲骨系列

红羽飘摇

2015/2016 AUTUMN/WINTER STYLE SHANGHAI FASHION TREND

[海派自然风格]
SHANGHAI NATURAL STYLE

sustain 续

花团锦簇

一枝独秀

INTERPRETATION OF NATURAL BEAUTY BY ORGANIC FORMS

In Shanghai style, such natural style and processing can be continued naturally, with the use of coquettish yet pure colors, to show the charm of natural origin as well as unlimited passion and imagination of designers. Interspersed with rich and varied structures, it presents the charmingly natural and dreamy colors by designers.

With the decline of natural resources and ecological environment, it is found that returning to nature is the future direction of learning. By using the flowers, feathers, cane and other natural elements, plus cord knitting, bamboo knitting and other technologies, it shows a unique sense of style. The combination of different craftsmanship and various structures in each works embody the exploration and pursuit to the beauty by designers.

万艳齐飞系列

2015/2016秋冬海派时尚流行趋势

续 sustain
[海派自然风格] SHANGHAI NATURAL STYLE

沈沉
Shen Chen
水舞深造

东华大学服装·艺术设计学院
纺织品艺术设计专业教研室主任
上海水舞深造文化传播有限公司创意总监
国家纺织产品开发基地·特聘评估专家

蔓延于自然的时尚憧憬

这是一个在当代电影、广告等媒体眼中充满奇异的主题。卷草纹样郁郁葱葱；山花烂漫在霓虹灯下；蔓藤植物攀附在玻璃幕墙上；鲜花、或者被记忆的灰尘笼罩散发着工业制品的光泽。提花凹凸软硬对比的表面；规则与不规则的起泡轧花、平整与局部的压皱针织物；自然与工业品的混色效果；呈现出的种种奇异都来源于导演、艺术家及设计师对自然的憧憬。

城市中的人们强烈渴望对自然的亲近与探索。自然总是呈现出一种奇幻的人工理想般的面貌。对自然的疏远使得现实与幻想、真实与虚构、自然与非自然的界定变得模糊。科幻色彩、组织、质地、纹样等填充着现实世界的真实，寻求理想的平衡。

鼎天时尚 沈沉《牡丹亭系列-良宵》
298g/㎡，涤纶100%，梭织拔色印花磨毛洗

水舞深造 沈沉、狄晨光《奶油牡丹转基因》
251g/㎡，涤纶100%，梭织印花涂层

明阳集团 沈沉《骥美人转变》
228g/㎡，氨纶31%，羊毛55%，腈纶16%
金银丝8%，针织

2015/2016 AUTUMN/WINTER STYLE SHANGHAI FASHION TREND

[海派自然风格]
SHANGHAI NATURAL STYLE

sustain 续

EXPECTATION OF STYLE IN NATURE

This is the fantastic subject in the eyes of contemporary film, advertising and other media. The curled grass is green and lush; the flowers are blooming in neon; the vines cling to the glass curtain wall; the flowers in memory exude the sheen of industrial product. It is designed with the uneven contrast of soft and hard surfaces of jacquard; regular and irregular blistering ginning, with partial flat knitted fabrics; mixing effect of natural and industrial products. All these show that a variety of bizarre are derived from the visions of nature by directors, artists and designers.

The people in cities have a strong desire to get close to nature and exploration. The nature always presents the ideal look like a fantasy. The alienation to nature makes the blurred definition on reality and fantasy, truth and fiction, natural and unnatural world. The real world is filled with the fantastic color, organization, texture, pattern and other elements, to seek the ideal balance.

鼎天时尚 《红珊瑚巴洛克》
364g/㎡，棉15％，氨纶37％，羊毛48％，提花

水舞深造 沈沉、狄晨光《似水年华》
511g/㎡，棉100％，梭织印花

鼎天时尚 《红珊瑚色贵气花》
261g/㎡，涤纶40％，棉8％，金银丝19％，腈纶33％，提花

达利丝绸 沈沉《花开富贵》
380g/㎡，皮革100％，印花压模

2015/2016秋冬海派时尚流行趋势

续 sustain
[海派自然风格]
SHANGHAI NATURAL STYLE

倪明
Ni Ming

东华大学服装·艺术设计学院
纺织品艺术设计专业教师

时尚面料设计中的可持续主题

时尚艺术是生动的、富有感染力的，是具有大众性、交叉性的流行艺术。时尚既然离不开流行，也自然少不了传承。一股时尚的前卫潮流，代表着独特的社会文化及社会风尚，文化和风尚创造时尚。文化和思潮的变迁是沿着人类生活和发展的轨迹前行，新的时尚自然也在继承中孕育生发起来。同时可持续时尚也体现了人们在生活中保持关注的主题，如环境保护和绿色设计，随着时间的推移，其价值反而被不断挖掘出来。这也揭示了时尚和大众生活息息相关的规律，让每个人能从时尚中认识生活、认识社会。所以，持续性的时尚主题总是具有可以无限阐释的空间，随着大众的关注其价值依然存在并得到新的认同。

春天的雨季

有灌木的风景

古城印象

2015/2016 AUTUMN/WINTER STYLE SHANGHAI FASHION TREND

[海派自然风格] SHANGHAI NATURAL STYLE

sustain 续

潮湿空气中的房屋

凝结在空气中的芬芳

SUSTAINABLE FASHION THEME IN FABRIC DESIGN

The fashion art is a vivid popular art with infectivity, publicity and intersectionality. Since it is inseparable from the popular fashion, it is naturally inseparable from the inheritance. An avant-garde fashion trend represents the unique social culture, social habits and creative fashion. As the change of the culture and thought is moved ahead along the track of human life and development, the new fashion is naturally born in the breeding. Meanwhile the sustainable fashion also reflects the concerning theme in people's daily life, such as environmental protection and green design, whose value has been continuously excavated as time goes by. It also reveals the relevant rules of the fashion and public life so that everyone can have a understanding of the life and society from the fashion. Therefore, the sustainable theme of fashion always has unlimited space for interpretation and its value still exists with the public concern and will obtain a new identification.

2015/2016秋冬海派时尚流行趋势

续 sustain
[海派自然风格] SHANGHAI NATURAL STYLE

许德民
Xu Demin

中国当代抽象艺术家
中国抽象艺术学学科创始人
中国抽象文化创导者
《中国抽象艺术》主编
国家一级美术师
复旦大学中文系特聘教授

创新水墨画的追求

水墨画的难题依旧是图式与风格问题。我认为抽象水墨绘画绝对是形式大于观念,因为形式代表永恒、代表纯粹。绝对的形式代表绝对的艺术,绝对的技术就是绝对的艺术!当水墨画的材料被限定之后,突破的唯一途径只能是技术。技术突破带动符号突破,实现图式风格突破,观念创新也就在其中了。因此,形式独创和唯一,就是我努力的方向。

我的创新水墨画追求的就是通过绝对偶发而获得的绝品、神品和逸品!通过独特的绘画技法,表现独特的中国水墨气质和魅力,营造体现中国艺术精神的符号象征。

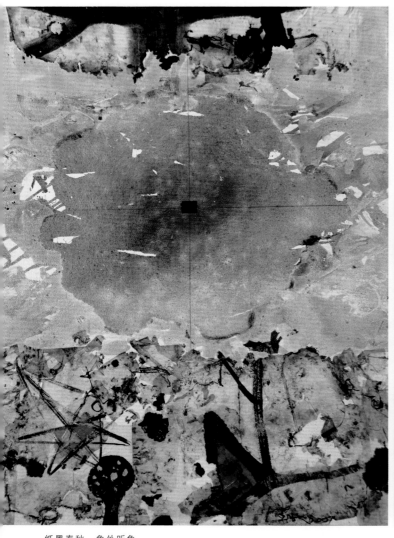

纸墨春秋·象外听象

纸墨春秋·思观形系

2015/2016 AUTUMN/WINTER STYLE SHANGHAI FASHION TREND

[SHANGHAI NATURAL STYLE] 海派自然风格

sustain 续

中国墨魂系列

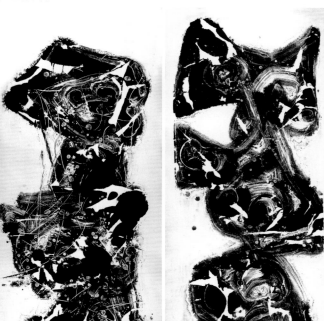
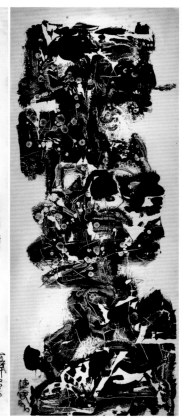

纸墨春秋·亡泊启湖

Hope to achieve the dream of artistic originality and uniqueness; Hope to make an equal dialogue between the global art and Western art; Hope to become an innovative coordinate in Chinese art! The problem for ink painting is still pattern and style. I think that form is definitely greater than idea for abstract ink painting, because the form is a representation of purity and eternality. The absolute form is a symbol of absolute art technology, and the absolute technology is absolute art! When the ink material is limited, the only way to break is the technology. The symbolic breakthrough is driven by technology breakthrough, to achieve the breakthrough of pattern and style, with concept innovation. Thus, the original and unique forms are the direction of my efforts.

My pursuit of innovative ink painting is to obtain the pure and great masterpiece in an absolutely incidental manner! Through the unique painting techniques, the unique temperament and charm of Chinese ink painting are presented, to create and embody the symbol for spirit of Chinese art.

PURSUIT OF INK PAINTING

2015/2016秋冬海派时尚流行趋势

续 sustain

[海派自然风格]
SHANGHAI NATURAL STYLE

首饰留下永恒自然美好

我们对自然的一切事物都包含情感，以自然之形态衍生出了无数优秀的作品。自然的形、自然材质和源于自然的手工技艺，"首饰"能以一些方式永远留下自然的美好。

《藕遇》首饰作品将白色的月光石用金属相连接，采用特别的包镶方式将宝石通透的质感表现出来，而造型灵感来源"藕"被切断后的自然形态。"妾心藕中丝，虽断犹连牵"，以此为灵感，表达了让人牵肠挂肚的爱情之美好。

《荷塘秋色》首饰作品以独特的形式美呈现了文人画的画面意境。作品主体形象是秋日残荷，以一种不让人怜悯却心生敬意的姿态，在生命尽头处呈现另一种美和精致。作品体现了一种积极的人生观。

傅婷
Fu Ting

东华大学服装·艺术设计学院
产品设计系教师

《藕遇》穿戴图之一

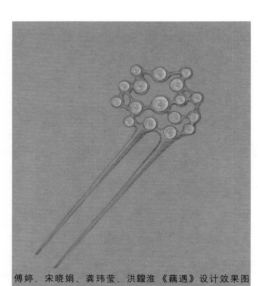

傅婷、宋晓娟、龚玮莹、洪鍠淮 《藕遇》设计效果图

《藕遇》灵感图

《藕遇》穿戴图之二

2015/2016 AUTUMN/WINTER STYLE SHANGHAI FASHION TREND

[海派自然风格]
SHANGHAI NATURAL STYLE

sustain 续

ETERNAL BEAUTY BY JEWELRY

We have the deep emotion to all things, and produce the numerous outstanding works in the natural forms. With the natural shapes, materials and craftsmanship, the natural beauty of "jewelry" can be cherished forever in some way.

The jewelry works of "Meeting with Lotus" connects the white moonstone with the metal, and uses a special covering to manifest the transparent texture of gem, while the styling is inspired by natural form of "lotus" after being cut. "My love, like a lotus root, linked by fibers though divided", it is an expression of beautiful love that people yearn for.

The jewelry works of "Autumn in Lotus Pond" presents a beautiful picture of literati painting in a unique form. Its main image is the withered lotus in autumn, with a pity but admirable gesture, showing another kind of beauty and sophistication at the end of life. The works reflects a positive outlook on life.

《荷塘秋色》穿戴图

傅婷、王敬宇 《荷塘秋色》设计效果图

《荷塘秋色》灵感图

2015/2016秋冬海派时尚流行趋势

趋势演绎 TREND INFERENCE

革新·生机 RENOVATION·VITALITY

趣 fun

[海派都市风格]
SHANGHAI URBAN STYLE

2015/2016秋冬海派时尚流行趋势主题
2015/2016 AUTUMN/WINTER STYLE SHANGHAI FASHION TREND THEMES

革新·生机　RENOVATION·VITALITY

2015/2016秋冬海派时尚流行趋势

主题说明 THEME DESCRIPTION

革新·生机 RENOVATION·VITALITY

趣 fun
[海派都市风格] SHANGHAI URBAN STYLE

趣
中文拼音：qù
英文：fun，interesting，fascinating
组词：乐趣，有趣，趣味
趋势分类：海派都市风格
趋势简述：年轻快乐的多彩都市潮流

  在2015/2016秋冬"革新·生机"的流行趋势主题下，海派都市风格所演绎的是散发都市年轻正能量的创造革新，充满了活力、生机与想象。
  海派都市风格是指在国际化大都市背景下，以充满趣味的波普、卡通、游戏、涂鸦为主要设计元素，演绎亮丽多彩，轻松快乐的海派都市精神，倡导青少年设计文化与新生设计力量。
  知名时装设计师李鸿雁的Helen Lee品牌代表都市摩登和现代东方时尚，用设计师独特的视觉、现代的设计去分享和互动品牌和当代中国人对生活的各种积极追求，反映新时代的上海和新上海人的时尚，其迪斯尼米老鼠系列时装设计得到了广泛关注。平面设计师虞惠卿的Shanghai Type动态字体秀是通过互联网形式传播的"城市与字体"的设计活动，为平面设计增添了趣味和创新。圆周率品牌传播机构执行创意总监倪海郡把煤饼、热水瓶、脚踏车、红宝书等老上海耳熟能详的物品进行了趣味演绎。大学生设计新锐们更是对充满玩味的设计要素情有独钟，创造出90后、00后热爱的新鲜玩意儿。充满童趣和儿时记忆的设计要素成为很多设计师的灵感来源，充满压力的紧张都市人日益热衷于这种能带来快乐的设计风格。

2015/2016 AUTUMN/WINTER STYLE SHANGHAI FASHION TREND

主题说明
THEME DESCRIPTION
革新·生机 RENOVATION·VITALITY

趣
fun
[海派都市风格]
SHANGHAI URBAN STYLE

Fun
Chinese Pinyin: qù
English: fun, interesting, fascinating
Group of words: pleasure, interesting, fun
Trend category: Urban Shanghai Style
Trend description: colorful urban trend of youth and happiness

Under the theme of the "Renovation · Vitality" fashion trend in the 2015/2016 autumn and winter, the Urban Shanghai Style demonstrates the creative reform with the positive energy of the urban young generation, which is full of vitality, vigor and imagination.

The Urban Shanghai Style refers to taking the interesting bophut, cartoons, games and graffiti as the main design elements to interpret the bright and colorful, relaxed and happy urban Shanghai Style spirit, which is to advocate the teenagers to design the culture and the designing forces of the youth under the background of the international metropolis.

The Helen Lee brand of the well-known fashion designer, Li Hongyan, represents the modern urban fashion and modern oriental fashion, which uses the unique visual of the designer and the modern design to share and interact with the brand and various active pursuits of life for the modern Chinese people, reflecting the Shanghai in a new era and the new fashion of Shanghai people. The fashion design of Disney Mickey Mouse series has also been widely concerned by people. The "Shanghai Type" dynamic fonts show of the graphic designer, Yu Huiqin, is a designing activity to spread the "city and font" via the Internet, which has added the interesting and innovation to the graphic design. Ni Haijun, the design director of the girth quotient brand communication agency, has made an interesting interpretation on the briquette, thermos, bicycles, Little Red Book and other familiar items in the old Shanghai. The cutting-edge university designers tend to be more focused on the interesting design elements, creating the fresh stuff loved by "the generation after 90s" and "the generation after 00s". The design elements full of children's interesting and the childhood memories have become the source of inspiration for many designers. People living in the city who are faced with stress and pressure are gradually keen on this happy design style.

2015/2016秋冬海派时尚流行趋势

趋势演绎
TREND INFERENCE

革新·生机 RENOVATION·VITALITY

趣 fun
[海派都市风格]
SHANGHAI URBAN STYLE

不拘一格的街头时尚为本系列带来灵感。多种色彩的交汇在混乱的重组中建立新的秩序。春夏季的酸性色与粉蜡色继续延续，蓝绿色调在这一季更加的饱和。

The eclectic street fashion brings the inspiration for this series. The intersection of various colors creates a new order in the reorganization of chaos. The acidic and pastel colors of spring and summer continue to follow, and the blue green hues become more saturated in this season.

2015/2016 AUTUMN/WINTER STYLE SHANGHAI FASHION TREND

主题色彩 COLOR THEME

蜜橘橙 SH0202074	棉花白 SH0800405
金币黄 SH0101053	魅丽蓝 SH0600341
青柠绿 SH0903824	深靛蓝 SH0501024
水泥灰 SH0401967	午夜黑 SH1000080
松石绿 SH0601122	孔雀蓝 SH0503009
电光蓝 SH0504137	杜鹃红 SH0302994
蓝莲紫 SH0402289	泰迪黄 SH0101436
浅玫红 SH0301194	烟熏紫 SH0402699
玫瑰红 SH0400221	葡萄紫 SH0402289
泳池蓝 SH0600401	烈焰红 SH0300166

2015/2016秋冬海派时尚流行趋势

趣 fun
[海派都市风格]
SHANGHAI URBAN STYLE

快乐的摩登东方时尚

HELEN LEE品牌以充满活力的城市——上海为基地，设计元素都以回眸上海的过去，拥有的现在，想象中的将来，运用现代的设计灵感去提炼出品牌的文化，撇弃和打破刻意拼贴中国传统元素才是中国设计的观念，设计代表都市摩登、现代东方时尚。品牌的宗旨理念是：用设计师独特的视觉、现代的设计去分享和互动品牌和当代中国人对生活的各种积极追求，反映新时代的上海和新上海人的时尚，用时尚设计去体现优雅、品味、摩登、积极和创造性并存的身份象征。

我一直以为时尚是随着人们的生活方式和节奏来的。现在大家都想有一个轻松和欢乐的环境，这些卡通形象已经非常经典，在人们心中有着根深蒂固的印象，把一个经典卡通放在时尚里面就看我们怎么去搭配和表达。2015/2016秋冬海派时尚流行趋势在充满想象力的同时要能够把这样的卡通形象传播出去，我想这不仅仅是一个卡通人物，它也可以更加时髦和时尚。

李鸿雁
Helen Lee

本土知名设计师
HELEN LEE品牌创始人

与迪士尼合作的"90+10"系列

"WAKE 唤醒"系列

2015/2016 AUTUMN/WINTER STYLE SHANGHAI FASHION TREND

[海派都市风格]
SHANGHAI URBAN STYLE
fun 趣

上海时装周2013秋冬女装

MODERN ORIENTAL FASHION WITH HAPPINESS

The HELEN LEE brand is based in the vibrant city – Shanghai, and all the design elements are to retrospect to Shanghai in the past, the owned present, the imagined future by using the modern design inspiration to refine the brand culture, which abandons and breaks out the concept that the deliberate Collage of Chinese traditional elements is the Chinese design with the representative of the modern city and the modern oriental fashion; The purpose and philosophy of the brand are: to share and interact the brand and the various active pursuits to the life in the contemporary China and to reflect the Shanghai in the new ear and the fashion of Shanghai people with her unique visual, modern design; to reflect the elegance, taste, modernism, activity and creativity with a coexistence of status symbol with her fashion and design.

I always thought that fashion is followed by the way and pace of people's lifestyle. Now everyone wants to have a relaxed and convivial environment and these cartoon characters are very classic that they have formed an ingrained images in people's minds. It is decided by us how to do with the match and expression by integrating a classic cartoon into the fashion. The fashion trend of 2015/16 autumn and winter should be able to spread out this cartoon with full imagination. I think this is not just a cartoon character and it should be more fashionable and stylish.

2015/2016秋冬海派时尚流行趋势

趣 fun
[海派都市风格] SHANGHAI URBAN STYLE

潘剑锋
Pan Jianfeng

跨文化革新设计机构-字研所(SHTYPE)
创作总监

设计的跨文化沟通与趣味

　　设计就像生命的另外一个载体一样,当你设计出一个东西来,你要感受到它的体温。从我们儿时就接触的江南文化的书法练习中,去结合现代西方的文明,开发出一些符合时尚、符合今天人们生活的东西。在上海当代艺术馆中让观众自由互动的涂鸦,使观众们可以参与其中,体验思考艺术和日常生活的一种趣味交互。

　　设计的跨文化沟通,不仅是国家和国家之间不同的文化沟通,其实我们每一个个体,都有不同的文化背景。中文字体如何从江南文化与西方文化的交互中得以创新?在今天这个时代,非常值得关注。我希望建立一个开放的系统,而不是单独的一个标志,因为单独的标志已经不能满足时下这种沟通的需要。

冲中苏字典

上海当代艺术馆 "新版飞行棋"

大口杯

THE CROSS-CULTURAL COMMUNICATION AND TASTE OF THE DESIGN

以中国书法为灵感的字体设计

The design is like another carrier of the life. When you are designing a product, you may feel its body temperature. Some products applying to the fashion and today's life should be developed with an integration of the modern Western civilization based on the calligraphy practice of Jiangnan culture in our childhood. The free interactive graffiti with the audiences in Shanghai Museum of Contemporary Art got the audiences involved in it to experience and think about the interesting interaction between the art and the daily life.

The cross-cultural communication of the design is not only to have different cultures between the countries but each of us as an individual shall have some different cultures. How can the Chinese words be innovated from the interactive exchange of the Jiangnan and western cultures? It is very worthy of attention in this day and age. I want to create an open system, rather than a separate symbol because the single symbol cannot meet needs of today's communication.

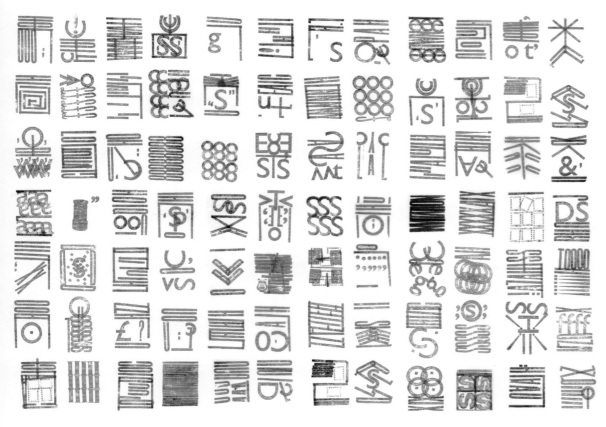

英国英格兰中央大学伯明翰艺术设计学院画廊"长"

外滩十八号"圣诞'倒了'"

2015/2016秋冬海派时尚流行趋势

[海派都市风格]
SHANGHAI URBAN STYLE

平面设计的新媒体进化
「ShanghaiType」动态字体秀

THE NEW MEDIA REVOLUTION OF GRAPHIC DESIGN

作为集结了设计艺术与数字、信息技术为一体的新媒体艺术，其在平面设计中的应用已从理论走向了实践，为平面设计的创新与发展注入了新鲜的血液。新媒体艺术的加入，丰富了平面设计的视觉语言，增进了平面设计方法的多元化、扩展了平面设计的发展空间，并且实现了平面设计的信息可视化。当下与未来，新媒体艺术必将引领平面设计的发展到达更高的维度，走向更新的未来。

虞惠卿
Yu Huiqin

平面设计师
时浪Snap设计合伙人兼创意总监
「ShanghaiType」动态字体秀 策划
「纸介」艺术设计邀请展 策展
「平面幸福-荷兰平面设计百年展」邀请策展
「我爱上海」字体设计邀请展 策划

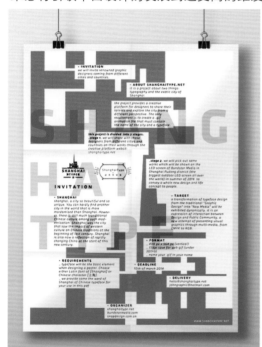

"ShanghaiType"动态字体秀－是通过互联网自媒体形式传播的"城市与字体"的设计活动。将Facebook、Behance、Weibo、WeChat等等自媒体平台链接起来，让不同地域的设计师互动和人士。本次活动邀请了来自17个国家的32位国际设计师以及46位国内新锐设计师，共计86幅动态设计作品。

活动主题包含两个文化元素：字体-城市以"SHANGHAI""上海"为字体设计基本元素用动态视觉形式展示来自不同城市和国家的设计对上海的理解和感知。

观看动态字体请登录：
ShanghaiType.net

86　www.style.sh.cn

2015/2016 AUTUMN/WINTER STYLE SHANGHAI FASHION TREND

About ShanghaiType Motion Graphics Show

It is a project about two things: typography and the exotic city of Shanghai.

The project provides a creative platform for designers to share their talents and explore the city from a different perspective. The only requirement is to create a .gif animation file that must contain the name of the city and a typeface.

This project is divided into 2 stages.
Stage 1: We will share with those designers from different cities and countries on their works through the creative platform websit: shanghaitype.net
Stage 2: We will pick out some works which will be shown on the LED screen of Bundstar Media in Shanghai Pudong district (the biggest outdoor LED screen all over the world) in summer of 2014 to convey a whole new design and life concept to people.

Understand Sociology, Not Technology

We will invite renowned graphic designers, who come from different cities and countries.
A transformation of typeface design from the traditional "Graphic Design" into "New Media" will be exhibited dynamically. It is an expression of interaction between Design and Public Community, a new attempt of presenting visual graphics through multi-media, from CMYK to RGB.

案例分析

MOTION GRAPHICS
CMYK > RGB

将字体设计从传统的"平面设计"转换到"新媒体"的动态展示。探索设计与公共空间互动的表达，从CMYK到RGB的模式转换，以及通过多媒体形式展示视觉图形的新尝试。

Götz Gramlich / gggrafik_德国

Lopetz-Büro Destruct_瑞士

Felix Pfäffli_瑞士

余子骥_深圳_中国

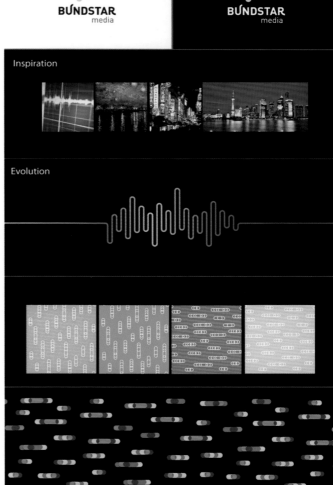

动态LOGO

观看动态LOGO

2015/2016秋冬海派时尚流行趋势

倪海郡
Nathan Nee

上海圆周率品牌传播机构合伙人兼执行创意总监
德国汉堡品牌学院客座教授
上海视觉艺术学院客座教授
上海市广告协会年度优秀作品评审 委员会专家
上海市广告协会设计 委员会委员
上海创意工作者协会 会员
上海市黄浦区人才协会 会员
海峡两岸大学生设计艺术节 评委

妙趣横生老上海

煤饼、热水瓶、红宝书等老上海耳熟能详的经典物品在这里进行了趣味演绎。曾经作为国人奢侈品的凤凰牌缝纫机、永久牌自行车，如果贴上西方限量版奢侈品的标签将会怎样？红宝书中的经典语句在当代海派语境下又是何种创新释义？

沪上设计相对更务实，讲究精致独立，亦中亦西。无门无派反倒独树一帜。时尚在上海更加尊重商业，急急忙忙，倒也井然有序。这里有中国最时尚、精致、独立的女性，快节奏生活让很多人望而却步，沉淀下来的自然是精华！

紅寶書

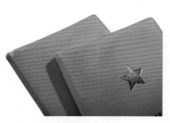

2015/2016 AUTUMN/WINTER STYLE SHANGHAI FASHION TREND

[海派都市风格] SHANGHAI URBAN STYLE
fun 趣

煤饼

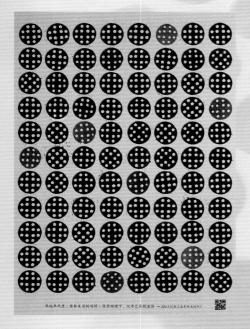

THE OLD SHANGHAI FULL OF WIT AND HUMOR

The interesting interpretation has been here made to the briquette, thermos, little red book and other old familiar classic items in Shanghai. What would it be if the Phoenix sewing machines and the "Forever" bicycle that used to be the oriental luxury brands are marked with the Western limited edition luxury? What would be the innovative interpretation of the classic statement in the little red book in the context of contemporary Shanghai style?

In Shanghai, it is a relatively more pragmatic design, which focuses on the eastern and western delicacy and independence. No particular sect is actually unique. Fashion is more respected in the business here, which is in a hurry but everything is in good order. There are China's most stylish, sophisticated, independent women here and the fast-paced life makes many people step back so that the precipitated is the essence.

快乐是什么—老人篇

快乐是什么—白领篇

大白兔内衬设计

大白兔内衬设计

大白兔奶糖

2015/2016秋冬海派时尚流行趋势

趣 fun
[海派都市风格] SHANGHAI URBAN STYLE

achette
上海雅氏鞋业有限公司

个性设计重塑，创造与乐活并行

自然、简洁、乐活是法国雅氏（achette）品牌始终秉承与倡导的品牌理念。多年来，雅氏在女鞋设计上，与时俱进，不断突破创新，融合独立、自由、充满个性化的元素，时而激进，时而怪诞，无一例外地表现创造性，却又保持了经典法国品牌的优雅气质。

对于2015/2016年秋冬季，流行趋势聚焦在重新塑造独具个性的形态上，通过不规则几何设计、色彩的强烈对比、街头元素的结合以及特殊工艺和材质的运用来进行诠释。雅氏将出色的廓型和独特的风格作为本季度的设计导向，使女性在享受舒适的同时，以品牌新定位的时尚态度，积极乐观地面对充满挑战的现代生活。

"albip"鞋子

"bilbon"鞋子

"bilboa"鞋子

2015/2016 AUTUMN/WINTER STYLE SHANGHAI FASHION TREND

THE REMODELING OF THE INDIVIDUAL DESIGN, PARALLELISM WITH CREATIVITY AND LOHAS

The natural simple LOHAS has always been the concept of the brand advocated by the French Jakob (achette). Adhering to this philosophy, Jakob (achette) has advanced with the times for many years to integrate with more independent, free and personalized elements with constant breakthroughs in the innovation. The women's shoe design made by Jakob (achette) is sometimes radical and weird, but no one is not full of creativity. In the 2015/2016 autumn and winter, the fashion trends were focusing on remodeling a personalized form, whose performances were: irregular geometry, the concentrated expression of color, street elements and the use of special materials.

纯乳胶鞋底

2015/2016秋冬海派时尚流行趋势

趣 fun
[海派都市风格] SHANGHAI URBAN STYLE

刘乐
Liu Le Carany 卡拉羊

卡拉羊品牌设计师

张帅
Zhang Shuai CONWOOD

CONWOOD品牌设计师

高平华
Gao Pinhua 新秀集团 NEWCOMER GROUP

新秀品牌设计总监

绚彩多色旅游箱包系列

青春、活泼的气息是卡拉羊品牌最突出的理念，大胆的色彩搭配总是紧跟甚至引领时代的潮流。（为我们美的世界不断的增添美的元素，那种只属于"I'm young"般的激情，背上"卡拉羊"一起向快乐出彩。）

多彩乐高背包系列

www.style.sh.cn

2015/2016 AUTUMN/WINTER STYLE SHANGHAI FASHION TREND

箱包设计的年轻化、多彩化趋势

YOUNG AND COLORFUL DEVELOPMENT FOR BAG DESIGN

一直以来箱包都是时尚的宠儿，不管是日常生活还是聚会、旅行，你都需要用它来凸显自我个性、彰显独特风格、散发时尚魅力。现在的都市精神，活力四射、包罗万象，箱包的发展也越来越趋向于年轻化和多彩化，人们在追求健康、时尚、环保的生活方式的同时，也催生着更具设计感和青春活力的箱包。

2015/2016年海派时尚流行趋势的研究着重于对箱包色彩、形式和理念的研究。在设计上呈现大胆鲜艳的色彩及图案，欢乐玩乐的特质呈现在箱包上，反对正式呆板及高时尚的式样，同时考虑大众品味，为箱包发展增添更多盎然生机。

Bag has been the darling of fashion, highlighting their own individuality, unique style and stylish charm, regardless of daily life or party and travel. At present, the spirit of city is vibrant and inclusive, so the development of bag is in a colorful and young tendency. In the pursuit of healthy, fashionable and eco-friendly lifestyle, the bags in more youthful sense of design are also produced.

In 2015/2016, the research on Shanghai trends will focus on the color, form and concept of bags. The bags will be designed with bold and bright colors and patterns full of joyful fun, without rigid and formal high-fashion elements, while integrating the public tastes, to add much dazzling vigor to the bag development.

休闲运动背包系列

新秀箱包以其"环保、时尚、商务、休闲、旅行"的设计理念与宗旨，致力于开发具有时代文化内涵、美观实用的箱包系列产品。

CONWOOD年轻、时尚、活力的代名词。本着对旅行潮流执着探索的创意精神，为旅行爱好者打造时尚与功能兼备的高性价比旅行产品，和CONWOOD一起旅行去吧！（不从众不随波，以自己独有的方式，走自己的路，欣赏自己的风景。）

活力撞色背包系列

2015/2016秋冬海派时尚流行趋势

趣 fun
[海派都市风格]
SHANGHAI URBAN STYLE

灵感图片

沈沉
Shen Chen

水舞深造

东华大学服装·艺术设计学院
纺织品艺术设计专业教研室主任
上海水舞深造文化传播有限公司创意总监
国家纺织产品开发基地·特聘评估专家

在矛盾中感受设计趣味

不惊不诧、实已演变成怦然心动却平静如水。各种波纹大胆曲折的尝试和兼容了契合当下年轻消费群的特征：体验真实的心跳与面不改色的酷。更物质化的性感，不论是色彩还是造型，都因为真实感的增强而显得更加振奋。

这是一个成长矛盾的主题，鲜明生动又矜持不凡。既有整洁细致的一面，幼稚的图案配色老道沉稳；也有仿真造型所带来的不一样的前卫感，效果流畅的线条印花在后整理疏松或随意褶皱织物上。面料风格拼接转呈相悖并置趋势。

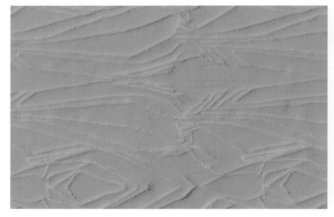
鼎天时尚《5A》
364g/㎡，涤纶75%，氨纶25%，梭织提花

ZOOUZA 沈沉《Pv国际展总监到访东华大学》
764g/㎡，棉15%，氨纶37%，羊毛48%，梭织针织复合

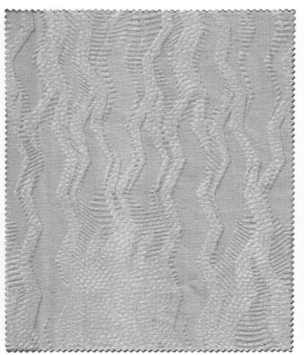
明阳集团 沈沉《汶川歌谣渐变》
598g/㎡，涤纶40%，棉40%，黏胶20%，双层复合

水舞深造 沈沉 狄晨光《未来一时空》
425g/㎡，涤纶色织起皱

2015/2016 AUTUMN/WINTER STYLE SHANGHAI FASHION TREND

[海派都市风格]
SHANGHAI URBAN STYLE
趣 fun

TO EXPERIENCE THE INTERESTING OF THE DESIGN IN THE CONTRADICTION

The unsurprised feeling has been actually evolved into the exciting yet calm emotion. The bold application of various corrugated twists is compatible with the characteristics of current younger consumers: cool and impassive experience of the true heartbeat. The more materialistic sexy feature will be more exciting due to enhanced realism, regardless of color or shape.

This is a theme of growing contradiction, vivid yet extraordinary. With neat and meticulous feature, the childish pattern is mixed with mature style; with the unique sense of avant-garde styling due to simulation, the smooth lines are casually printed on the corrugated fabrics. The fabric style stitching presents the contrary trend of juxtaposition.

达利丝绸 沈沉 傅鹏瑾《窗内窗外的风景》
204g/㎡，涤纶100%，转印花涂层

达利丝绸 沈沉《八卦基本形》
464g/㎡，棉48%，氨纶37%，羊毛15%，梭织印花

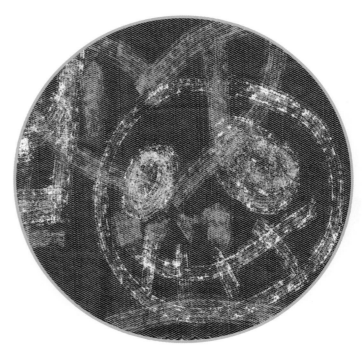

水舞深造 沈沉 陈梦琦《艺术家——可爱的骷髅们》
455g/㎡，棉100%，染色拔色印花

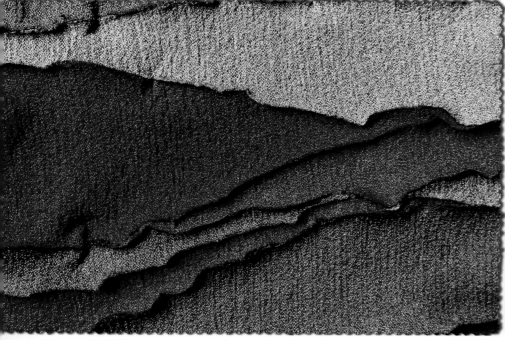

水舞深造 沈沉 王洁菲《片岩》
94g/㎡，棉25%，羊毛75%，针织拼合

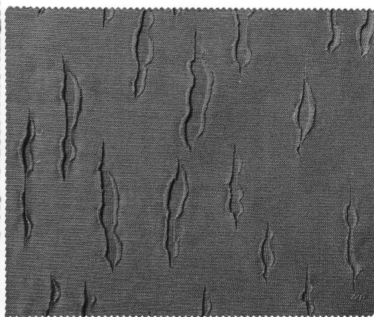

水舞深造 沈沉 顾颂恩《玛丽莲》
584g/㎡，棉15%，羊毛75%，
针织复合

2015/2016秋冬海派时尚流行趋势

趣 fun
[海派都市风格] SHANGHAI URBAN STYLE

倪明
Ni Ming

东华大学服装·艺术设计学院
纺织品艺术设计专业教师

男孩和女孩

趣味性是返璞归真的艺术写照

趣味存在于不同类型的设计产品中，以时尚、新鲜、轻松、幽默、流行为特征成为受人喜爱的审美类型，常常带给人们耳目一新的感觉以及令人心情愉悦。设计中的趣味性因素体现在从审美到创新设计方法的各个层面，趣味性鼓励设计师积极探索新思维、新方法、新媒介。趣味伴随着活力和创造力，无论是原始人和儿童的涂鸦，还是现代艺术大师的作品，其独特的视觉冲击力，大胆的用笔用色，富有想象的创造力，以及不可抗拒的感染力，都体现了浓厚的趣味性。趣味的难题是如何避免矫情和哗众取宠，而在设计中体现出质朴纯真的美，趣味性是创作者返璞归真艺术心灵的直接观照，是艺术回归自然、以人为本的现代时尚设计理念的外化表现。

海面之一

海面之二

2015/2016 AUTUMN/WINTER STYLE SHANGHAI FASHION TREND

苏格兰格子变奏图之一

苏格兰格子变奏图之二

INTERESTING DESIGN IS BACK TO BASICS

The interest exists in different types of product design. The aesthetic types characterized in the fashion, freshness, relaxation, humor and popularity to be populated by people usually brings a delighted fresh feeling to people. The interesting elements in the design are reflected in all aspects of design from an aesthetic approach to innovation and the designers are encouraged by them to actively explore the new ideas, new methods and new media. The interestingness is accompanied by dynamism and creativity. No matter it is the graffiti of the primitive people and children or the works of the modern art masters, its unique visual impact, bold pen colors, imaginative creativity and irresistible appealing are all reflected with a strong sense of interestingness. The challenges for the interestingness are how to avoid the hypocritical and grandstanding and how to reflect the pure beauty in the design. The interestingness is the direct contemplation of the creator's pure artistic mind and the external performance of the people-oriented design concept of modern fashion.

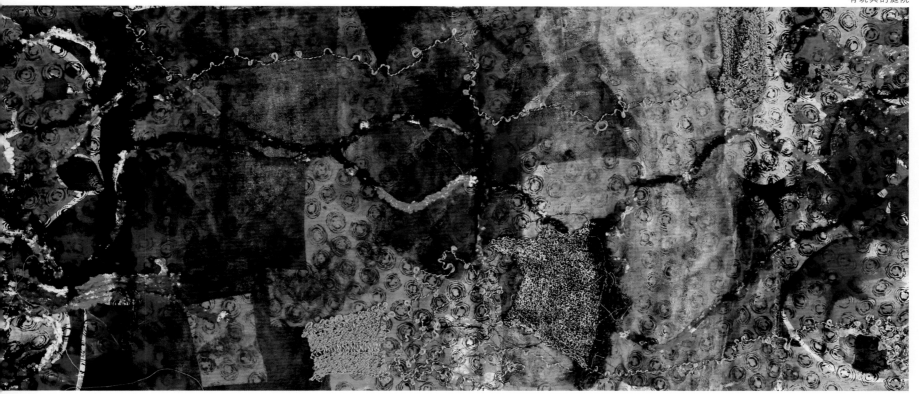

有玩具的庭院

2015/2016秋冬海派时尚流行趋势

趣 fun
[海派都市风格] SHANGHAI URBAN STYLE

傅婷
Fu Ting

东华大学服装·艺术设计学院
产品设计系教师

玩乐主义风潮下的小首饰

都市中流行的符号和事物，在现代感、玩乐主义、追求个性的风潮下，打造了绚烂多姿的首饰设计舞台，首饰的服饰化彰显了都市人的生活态度。

创可贴原本是人们常用的生活用品。作品通过再设计，赋予日常用品时尚的功能，它也可以是彰显个性、点亮生活的小装饰。该系列的设计有个性图案、可换色、能弹动几个特点。

能够表达自己态度和心情的手语是我们肢体语言的一部分，而别针又是让生活轻松便利的小帮手，"手语别针"在保留了别针本身实用性的基础上，手语的造型让人真的忍不住想将它们别在衣服上。

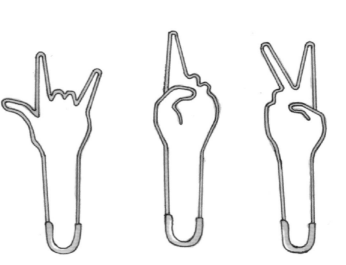

傅婷、方舒莹、王佩 《手语别针》效果图

《手语别针》佩戴图

《手语别针》灵感图之一

《手语别针》灵感图之二

2015/2016 AUTUMN/WINTER STYLE SHANGHAI FASHION TREND

A HIGHLIGHT OF LIFE ATTITUDE BY FASHION DESIGN OF JEWELRY

The urban popular symbols and things create a gorgeous stage of jewelry design, in the pursuit of modernity, hedonism and personality. The fashion design of jewelry highlights the urban people's attitude towards life.

The band-aid is a necessary household item in daily life. The works is redesigned to give the fashionable features to the household item, and also can be a small decorative item for highlighting the personality and lighting the life. The design has the personality patterns, changeable colors and dynamic features.

The sign language to express attitudes and feelings is part of our body language, and the pin is a helper to make life easy and convenient. The works of "Sign Language & Clip" retains the practical feature of pin itself, with the lovely shape of sign language that people really want to wear on the clothes.

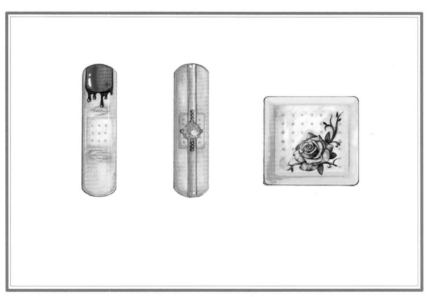

傅婷、龚玮莹《时尚创可贴》设计效果图

《时尚创可贴》穿戴效果图

《时尚创可贴》灵感图

   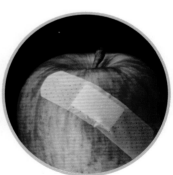

2015/2016秋冬海派时尚流行趋势

[海派都市风格]
SHANGHAI URBAN STYLE

吴亮
Wu Liang

东华大学服装·艺术设计学院
视觉传达系副主任

在玩味中寻求释放

在城市空间、美食、运动、音乐中每个人内心的向往和期待是什么?如果都市生活玄妙不可言喻,唯有融入它亲身去体验才可以知晓。无论是职场人对工作压力的释放,还是年轻人对自身活力的张扬,他们都应拥有自己的心理空间和物理空间。

杨少宸

王佳绮

时湾设计　　　田谜

2015/2016 AUTUMN/WINTER STYLE SHANGHAI FASHION TREND

RELEASE IN FUN

What is everyone's desire and expectation among the urban space, delicious food, sports, music imagination? If the urban life is ingenious beyond description, the only way to get to know it is to experience it by oneself. No matter it is the work pressure released by the career people or the vitality expressed by the young people, they should have their own mental space and physical space.

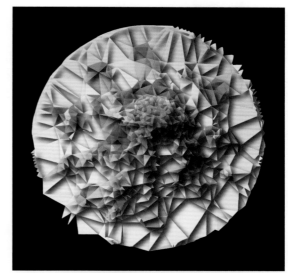

刘梦歆

张宪

陈思静

2015/2016秋冬海派时尚流行趋势

"趣"看上海

赵蔚
Zhao Wei

东华大学服装·艺术设计学院
视觉传达系教师

　　去年在纽约，我做了一个"另眼看纽约"的主题，基于我在上海生活的背景，记录在纽约让我驻足思考的各个片段。两座城市非常像，但不同的文化根源对思维认知的影响，让我捕捉到很多有趣的发现。我相信基于不同的背景经历和情感体验，在每个人的心中都有一个完全不同的上海。

　　我欣喜于年轻的学生们讲述给我关于上海的不同的精彩故事，他们有的认为上海的独特文化是一种"自嘲"，有的认为是一种"基因的突变"，有的认为是一种"悖论的美丽"，有的认为是一种"逆生长"，有的认为是杂糅的"甜味"，有的认为是能量转换的"两极"，是"拼图"，是"融汇"，是"无限"……

　　应该只有上海，才能给予我们如此纷繁又独特的感悟和灵感，这应该也是海派时尚的魅力所在。

王兆怡、潘姿燕、全莹、黄军衔 《甜》

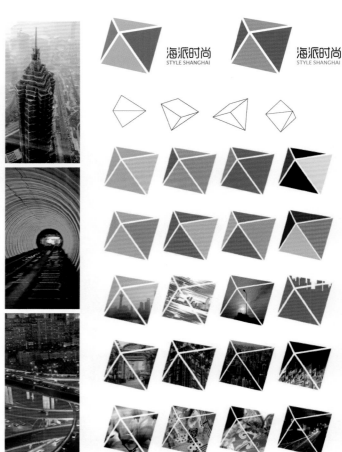

刘萱、戚铭、黄梦珍、胡妍琦 《多棱镜》

2015/2016 AUTUMN/WINTER STYLE SHANGHAI FASHION TREND

[海派都市风格]
SHANGHAI URBAN STYLE
fun 趣

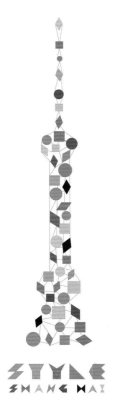

杭婧、王奥、陆艳阳、林灵 《碰撞》

INTERESTING LOOK AT SHANGHAI

Last year in New York, I did a theme of "seeing another New York", based on the background of my life in Shanghai to record the individual fragments that made me stop and think in New York. The two cities are much alike but I caught a lot of interesting discoveries about the impact on the cognitive thinking made by different cultural origins. I believe that there is a completely different Shanghai for everyone based on the different background experiences and emotional experience.

I am delighted at the different wonderful stories of Shanghai told by the young students. Some of them think that Shanghai's unique culture is a " self-mockery" ;some of them think that it is a "genetic mutation"; a "beauty of paradox"; a "reverse growth"; a mingled "sweetness"; the "bipolar" of the energy conversion, the "Jigsaw" ,the "integration " and "infinity"…

It should be only Shanghai that can give us numerous and unique insights and inspiration, which should also be charming of the Shanghai Style Fashion.

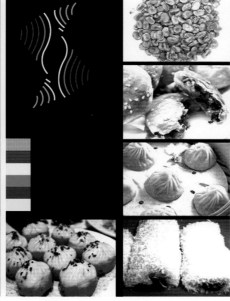

沈馨竹、颜斯圣、陈洁莉、王巧艺 《逆生长》

2015/2016秋冬海派时尚流行趋势

趋势演绎 TREND INFERENCE

革新·生机 RENOVATION·VITALITY

变 variety

[海派未来风格]
SHANGHAI FUTURISTIC STYLE

2015/2016秋冬海派时尚流行趋势主题
2015/2016 AUTUMN/WINTER STYLE SHANGHAI FASHION TREND THEMES

革新·生机 RENOVATION·VITALITY

2015/2016秋冬海派时尚流行趋势

主题说明 THEME DESCRIPTION

革新·生机 RENOVATION·VITALITY

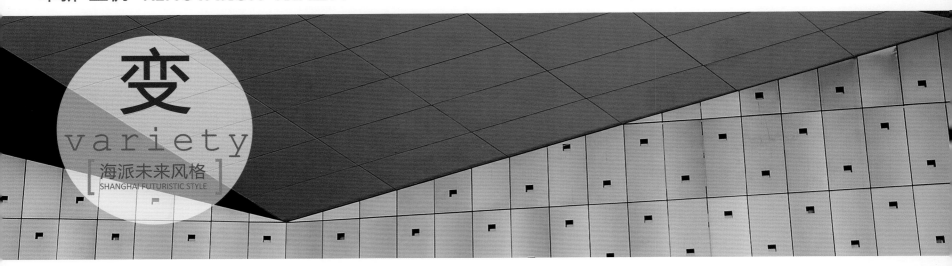

变 variety [海派未来风格 SHANGHAI FUTURISTIC STYLE]

变
中文拼音：biàn
英文：variety, change, transform
组词：变化，变形，变革
趋势分类：海派未来风格
趋势简述：高科技时尚设计变革

  在2015/2016秋冬"革新·生机"的流行趋势主题下，海派未来风格所演绎的是提升生活便捷的时尚高科技革新，带来时尚与科技跨界创意所迸发的生机和商机。

  海派未来风格是指具有未来风貌的海派时尚设计，新造型、新材料、新功能是关键。由3D打印技术引发的新型外观，由数码印花技术流行的炫彩图案，由面料科技带来的高功能面料，由可穿戴设备带来的服饰新功能，未来时尚在科技的引领下变得一切皆有可能。

  ACIA亚洲创意产业联盟创始人兼主席乔治·布迪漫认为亚洲创意设计产业以创新模式创造了可持续性的新商业价值。LKK集团首席设计官连振认为"科技+体验+时尚"将会是重要的时尚设计趋势，智能物联是一个重要的技术革新，而可穿戴设备是下一代重要的时尚产品。帽饰设计师BREMEN WONG采用PVC等新材料设计帽饰，希望人们能感受到形体以外的动态。海派未来风格提倡科技、人、自然三者达到设计的平衡，其根本是围绕"人"这一永恒的主题，做到人性化的设计。高科技所带来的时尚变革正在改变着人们的生活方式，也改变着设计师们的思维。

2015/2016 AUTUMN/WINTER STYLE SHANGHAI FASHION TREND

主题说明
THEME DESCRIPTION
革新·生机 RENOVATION·VITALITY

变
variety
[海派未来风格]
SHANGHAI FUTURISTIC STYLE

Variety
Chinese Pinyin: biàn
English: variety, change, transform
Group of words: change, deformation, reform
Trend category: Future Shanghai Style
Trend description: high-tech fashion design reform

Under the theme of the "Renovation · Vitality" fashion trend in the 2015/2016 autumn and winter, the Future Shanghai Style demonstrates the high-tech innovation to bring more convenience for the life and the vitality and business opportunities brought by the cross-border cooperation of the fashion and technology. The Future Shanghai Style refers to the Shanghai Style Fashion design with a future outlook and the key points are the new style, new materials and new functions. The new appearance triggered by the 3D printing technology; the colorful patterns populated by the digital printing technology; the high-performance fabrics brought by the fabric technology; the new features brought by the wearable devices, no fashion is impossible under the guidance of technology in the future.

George Budiman, the founder and chairman of the ACIA Asian creative industries alliance, believes that the Asian creative design industry has created the sustainable new business value with the innovation model. Lian Zhen, the chief design officer of LKK Group, believes that "Technology + Experience + Fashion" will be an important trend in fashion design and the intelligent instrumentation will be an important technological innovation but the wearable devices are important fashion products for the next generation. The hat designer BREMEN WONG uses the PVC new materials to design the hat, expecting that people may feel the dynamic condition outside the body shape. The Future Shanghai Style advocates the balance with the technology, human and the nature and it is basically about the eternal theme of "people" to achieve the user-friendly design. The fashion changes brought by the high-tech are changing the way people live and the way of the designers' thinking.

2015/2016秋冬海派时尚流行趋势

趋势演绎 TREND INFERENCE

革新·生机 RENOVATION·VITALITY

变 variety
[海派未来风格]
SHANGHAI FUTURISTIC STYLE

清新的冰冻色系影响了未来系列的色彩基调。纯净的冰川蓝、水晶橙、冰冻紫等色彩为冷静的灰色系和酷炫的蓝黑色带来跳跃感。荧光黄与荧光绿在这个系列更加的柔和。

The fresh frozen color affects the tone of the future series. The pure glacier blue, crystal orange, purple and other colors gives the sense of jumping to the calm frozen gray and cool blue black. The fluorescent yellow and fluorescent green in this series are much gentle.

2015/2016 AUTUMN/WINTER STYLE SHANGHAI FASHION TREND

主题色彩
COLOR THEME

粉蜡黄 SH0101056	曜石黑 Sh1000210
荧光绿 SH0900911	海军蓝 SH0500302
冰川蓝 SH0601281	星空蓝 SH0500979
水晶粉 SH0201832	岩石灰 SH1100753
薄荷绿 SH0503461	神秘紫 SH0400207
铁皮灰 SH1001021	冰霜白 SH0401919
迷雾灰 SH1100762	冰冻紫 SH0402006
香槟黄 SH0100392	矿石灰 SH1234567
水晶橙 SH0202123	冬白色 SH1200318
魔力蓝 SH0502245	亚铜灰 SH0701261

2015/2016秋冬海派时尚流行趋势

变 variety
[海派未来风格]
SHANGHAI FUTURISTIC STYLE

连振
Lian Zhen

LKK集团首席设计官
LKK上海总经理

智能物联的时尚

在产品设计领域，时尚不仅是外观造型，更是技术、成本、工艺、市场定位。在我们看来时尚需要有科技含量，智能物联是一个重要的技术趋势，而时尚是一种乐于表现自我及个人价值观的精神。对于时尚圈而言，可穿戴设备是下一代时尚产品重要的发展趋势。

"科技+体验+时尚"将会是重要的时尚设计趋势。有了科技才能高端，有了体验才能大气，有了时尚才能上档次。时尚在未来是一个基本词，任何事情都离不开时尚。

科技是一种时尚，而不仅仅是美学层面的，他应该包含很多信息。智能物联的时尚，让我们去改变世界！

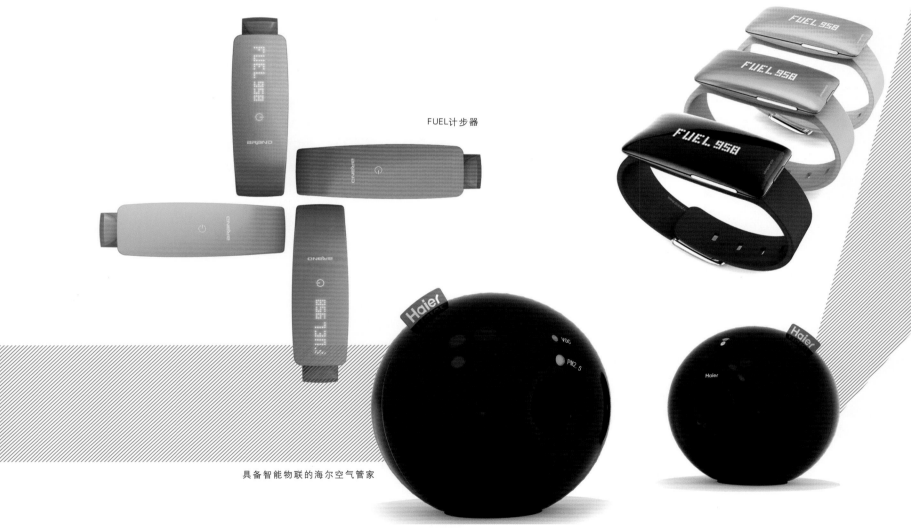

FUEL计步器

具备智能物联的海尔空气管家

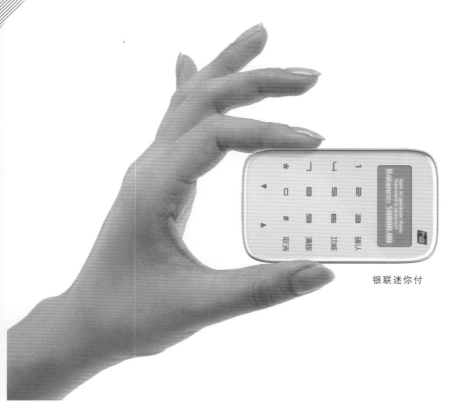

腕带手机

THE FASHION OF IOT

In the field of product design, fashion does not only means the appearance, but also for the technique, cost, technology and market positioning. In our opinion, the fashion needs to be scientific and technological and the intelligent instrumentation is an important technology trend but the fashion is a kind of spirit for people who are willing to express themselves and the personal values. For the fashion circles, the wearable devices are the important fashion trends for the generation of fashionable products.

"Technology + Experience + Fashion" will be an important trend for the fashion design. It will be high-end with the science and technology; it will be vigorous with the experience; it will be upscaling with the fashion. The fashion will be a common word in the future and everything is inseparable from fashion.

The science and technology is a kind of fashion rather than just for aesthetic level, which should include a lot of information. Let's go for the world change with the fashion of intelligent instrumentation!

银联迷你付

2015/2016秋冬海派时尚流行趋势

变 variety

[海派未来风格]
SHANGHAI FUTURISTIC STYLE

上海汤臣高尔夫别墅项目

乔治·布迪漫
George Budiman

ACIA亚洲创意产业联盟创始人兼主席
IDC(S)新加坡空间与室内设计协会会长
Design S新加坡创意产业联盟创意开发主席

亚洲灵感创造未来商业价值

随着亚洲的东方创意设计作品越来越受到世界欢迎,传统的设计产业也得到了新"人群"的再认识。这些具有特色的创意作品并不仅仅是对传统文化的抄袭,而是从深厚文化底蕴中发觉新的价值观后演变出的新时尚。以巧妙的商业创意模式包装后重新出发,创造可持续性的新价值。众所周知亚洲不可能永远作为跟风的创意基地,我觉得给亚洲不到10年时间,将可成为一个设计大洲。

现在不少西方著名品牌纷纷从东方文化中寻找设计新灵感。目前我所牵涉到的设计作品都会或多或少受到亚洲区域性文化的熏陶而得到灵感。尤其是借助正在没落中的传统手工艺进行创新和设计,既可以为保留传统手工艺做出贡献,对国际市场来说,也是具备吸引力的。上海必须要参与其中,作为一个富有深厚亚洲文化枢纽的国际大都市,上海在创意文化上具备了优势。创意设计产业的创新对于社会及经济增长的重要性与日俱增,我们都看到了这一领域的重要性,也注意到了未来创新和设计界会继续猛吹东方风的发展趋势。

马来西亚吉隆坡 luna酒吧

[海派未来风格]
SHANGHAI FUTURISTIC STYLE

variety 变

THE ASIAN INSPIRATION CREATES THE FUTURE BUSINESS VALUE

As the oriental creative design products in the Asia are becoming more and more popular in the world, the traditional design industry has also got a new recognition of new "people". These distinctive creative works are not just the plagiarism of the traditional culture, but they are the new fashion deriving from the discovery of the new values in the profound cultural background. It restarts after being decorated with a clever and creative business mode to create the sustainable new value. As we all know that Asia cannot always be the creative base that follows suit and I think the Asia will be able to become a design continent within less than 10 years.

新加坡裕廊 豪华套房

香港 The trillium

Now, many famous Western brands are looking for the new design inspiration from the oriental culture. The design works that I am currently involved in are more or less influenced by the Asian regional cultures to have the inspiration. It is specially the innovation and design by way of the declining traditional arts and crafts, which contributes to the preservation of traditional arts and crafts and also it is an attraction for the international market. As an international metropolis rich in the Asian culture, Shanghai must be involved in it and it has the creative and cultural advantages. The innovation of the creative design industry is becoming more and more important for the social and economic growth and we have seen the importance of this field but it is also noted the trend of the oriental culture will be continued in the future innovation and design sectors.

新加坡帕特森 豪华套房

2015/2016秋冬海派时尚流行趋势

变 variety
[海派未来风格]
SHANGHAI FUTURISTIC STYLE

徐燕辉
Xu Yanhui

MATCHBOX品牌创始人
兼设计总监

设计是提炼日常生活的状态

设计是提炼日常生活的状态，不是一些哗众取宠的概念。设计是为了美化生活，一个服装设计师的作品能让穿着者变得美好，就是最大的认可。

设计师是个幕后工作者，不需要用过于夸张的设计去博取众人的眼球证明自身的才华。你的作品，只需要让穿着者更美好。或许她有过浮夸的设计风格阶段，只是当下的徐燕辉状态很平静，不需要去做用力过度的设计去证明自己的与众不同。服装设计就是需要穿着者舒适，而不是让设计强大过穿着者的自身气场。

中国的传统材质与工艺特别需要被重视，从2011年开始，她和设计团队开始关注中国的传统材质的再运用，让传承至今的古老材质通过当代的设计语言去适应当下的生活方式，有新的生命力，星星之火，谁说没可能燎原，需要时间，沉淀。

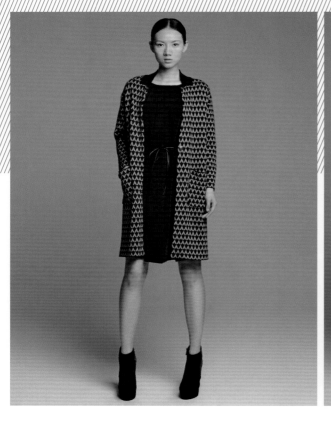

2015/2016 AUTUMN/WINTER STYLE SHANGHAI FASHION TREND

variety

[海派未来风格]
SHANGHAI FUTURISTIC STYLE

DESIGN IS REFINED FROM EVERYDAY LIFE

The design is to refine the state of daily life rather than some grandstanding concepts. It is designed to beautify the lives. It is the greatest recognition that a fashion designer's works make the wearers much better.

The designer is a worker behind the scenes, there is no need for whom to win everyone's eye to prove their talents with the exaggerated designs. Their works just need to make the wearers more beautiful. Perhaps she had experienced a flamboyant style stage and only the current state of Xu Yanhui is very calm without any overexerted designs to show her own distinctive design. The costume design is to make the wearers comfortable rather than to make that the design is over the wearer's own temperament.

China's traditional materials and crafts should be specially emphasized. Since in 2011, she and her design team began to focus on re-use of traditional materials in China to make the oldest materials inherited through the history to adapt to the current way of life by way of the contemporary designing language with new vitality. Who says that it is impossible for the single spark to start a prairie fire? It just needs time and precipitation.

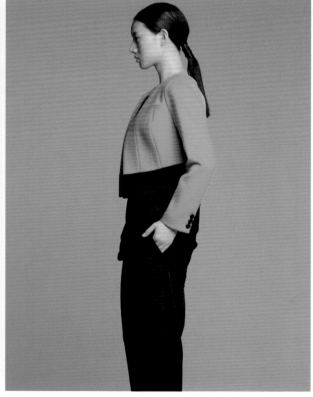

2015/2016秋冬海派时尚流行趋势

变 variety
[海派未来风格]
SHANGHAI FUTURISTIC STYLE

简单的新鲜

服装配饰品牌

VIVINIKO诞生在繁华而现代的大上海这块沃土，成立以来始终以"简单的新鲜感minium+"作为品牌设计理念，赋予极简主义以现代艺术感和街头文化之后所呈现出来的一种具有时代特征的新鲜感。用极简的手法，表现对新鲜感的理解。通过对"色彩、素材、体积"这三者关系的重新认识和塑造，并且在服装设计中融入对建筑学的理解，从而提供现代职业女性时髦而得体，实用而有艺术感的日常着装。"简单纯碎、廓型鲜明、功能主义"，具有显著的时代特征。

灵感来自对高科技印花技术的不断进步对自然界事物的新奇表现产生的思考，天白海蓝持续着生生不息的生命活力，对波浪的数码化变更融进时代美感，亲切如肌肤随身穿戴，无论是包袋还是鞋靴都选择舒适的款式，自然而又现代，简单而又新鲜。

海洋波纹数码印花系列手拎包/乐福鞋

2015/2016 AUTUMN/WINTER STYLE SHANGHAI FASHION TREND

[海派未来风格] 变
SHANGHAI FUTURISTIC STYLE

variety

海洋波纹数码印花双肩包

晚宴手拿包

男士系带乐福鞋

内增高休闲鞋

女士切尔西靴

SIMPLE FRESHNESS

    VIVINIKO was born in the prosperous and modern city of Shanghai. It has always adhered to "a simple freshness minium +" as a brand design philosophy to give the modern sense of minimalism in art and a freshness characteristic of the times is presented after the street culture. It expresses the understanding of the performance of freshness with a very simple approach. It provides the stylish and decent modern daily dress of practical and artistic sense for the career women through the new understanding and shaping of the relationships between "color, material, volume" and the understanding of architecture incorporated in fashion design. The "pure simplicity, sharp silhouette, functionalism" has significant characteristics of the times.

    It is inspired by the continuous progress of the high-tech flower printing technology, the thought on the new generation of new natural matters, endless vitality continued with the white sky and blue sea and the waving digital changes integrated into the era beauty. Both the bags and footwear are selected in a comfortable style with intimate wearing. It is natural and modern, simple and fresh.

2015/2016秋冬海派时尚流行趋势

变 variety
[海派未来风格]
SHANGHAI FUTURISTIC STYLE

东华大学–施华洛世奇
创意设计中心

闪耀的未来主义设计

施华洛世奇和连卡佛邀请东华大学学生运用施华洛世奇仿水晶元素，为全球最新最大的连卡佛上海旗舰店设计创意作品，学生作品于2013年底在连卡佛与国际著名设计师作品同台展出，成为施华洛世奇"时尚天桥"（RUNWAY ROCKS）的重要作品。

原闻、顾力文作品《ZEUS》用酷似雷电的三角形元素，打造出空间感和未来感；李芷玥作品《琥珀》采用亚克力数码喷印体现晶莹剔透的东方剪纸元素；马玉儒作品《"晶"灵》采用半透明塑料片组合出类似跳舞精灵般的感觉；杨楷浪、贺小令作品《重建》通过结构感的线条构建高科技世界。

李芷玥 《琥珀》

原闻、顾力文 《ZEUS》

杨楷浪、贺小令 《重建》

2015/2016 AUTUMN/WINTER STYLE SHANGHAI FASHION TREND

variety

SPARKLING FUTURISTIC DESIGN

The students from Donghua University were invited by the Swarovski and Crawford to design the creative works for the world's newest and largest Shanghai flagship store of Crawford by using the Swarovski crystal elements. The students' works are exhibited on the same stage with those of the internationally renowned designers in Crawford at the end of 2013, which have become the very important works of Swarovski (RUNWAY ROCKS).

The "ZEUS" works of Yuan Wen and Gu Liwen has created a sense of space and future with the triangle elements that are similar to the thunder and lightning; the "Amber" works of Li Zhi use the acrylic digital printing to reflect the sparkling oriental paper cutting elements; the "Crystal of Hope" works of Ma Yuru use the translucent plastic sheet to form a feeling like the dancing elves; the "Reconstruction" works of Yang Kailang and He Xiaoling build the high-tech world through the sense of structural lines.

马玉儒 《"晶"灵》

储思园 《打开我》

孙沁 《白雪公主的秘密》

2015/2016秋冬海派时尚流行趋势

变 variety
[海派未来风格]
SHANGHAI FUTURISTIC STYLE

未来帽饰中的感性创作

Bremen Wong
帽饰设计师

Bremen偏爱很大的头饰，因为觉得它们像个雕塑品，而这些发饰以不同材料组成，比如PVC、Cray、Acrylic等。假如说做夸张头饰是为了系列里有个代表作，倒不如说那是他能满足自己对头饰的渴望。他说："头饰不只是表面的点缀，也代表了一种生活态度。"

一件手工艺品，除了表面的细节和形态，是否有人真正去感受设计的意义？以纯熟的手艺创造出这些前卫夸张的头饰，他想要表达的是人应学习掌控自己的思想，弹性处理人生课题。将不好的事情给遗忘，把好的回忆锁在脑海。Bremen用他的心思与理念创造自己的作品，希望人们能感受到形体以外的动态。

Quarter 系列作品

创意头饰作品

Quarter 系列作品

2015/2016 AUTUMN/WINTER STYLE SHANGHAI FASHION TREND

[海派未来风格]
SHANGHAI FUTURISTIC STYLE

variety 变

Quarter 系列作品

EMOTIONAL CREATION IN FUTURE HATS

Bremen prefers to the big headwear due to their sculpture-like forms, and such headwear is made of different materials such as PVC, Cray and Acrylic.Rather than saying that the exaggerated headwear is to be a representative of the series, this is his achievement of desire to headwear.He said: "headwear is not only the surface embellishment, but also the representation of an attitude towards life."

Will the sense of design be appreciated for a handicraft works, in addition to its surface detail and form?These avant-garde and exaggerated headwear created with skillful craftsmanship reflects his thought that people should learn to control their own ideas and flexibly solve the life issues. To forget the bad things, but lock the good memories in the mind.Bremen creates his own works with his thoughts and ideas, and hopes that people can feel the dynamics outside the body.

2015/2016秋冬海派时尚流行趋势

变 variety
[海派未来风格]
SHANGHAI FUTURISTIC STYLE

朱卫明
Alexander Chu

时装设计师

年轻、自信与自由实验

　　Natural Gift是由品牌掌门人Alexander Chu带领充满活力、对时尚有独到见解的资深设计团队精心打造的一个充满个性、时尚、轻松的设计师品牌。Natural Gift为大都市的独立、自由、追求时尚和独特穿着品味的成功人士而设计，他们青春睿智、富有激情、重视风格更强调品质。对每一季的潮流进行解构重组，服装风格充满讽刺和反对随波逐流，体现自信、性感不失雅致的性格特征，并将高水准的剪裁融入其中。Natural Gift同时代表了一种生活方式：品牌诉说着年轻人的语言，以完全自由地试验材料和形状为乐。Natural Gift生活在地理边界的当代大都会，在音乐、科技、建筑等诸多领域中汲取刺激的灵感，然后转化成富有设计内涵的皮装系列。

2015/2016 AUTUMN/WINTER STYLE SHANGHAI FASHION TREND

[海派未来风格 SHANGHAI FUTURISTIC STYLE]

variety 变

THE YOUTH, SELF-CONFIDENCE AND FREEDOM TO EXPERIMENT

Natural Gift is a designer brand carefully created by the brand head Alexander Chu who led the vibrant design team with unique perspective on fashion, which is full of personality, fashion and relaxation. Natural Gift is designed for the successful persons with independence, freedom, the pursuant of fashion and unique dress sense, who are full of passion and youthful wisdom and focus on the style with more emphasis on the quality. It is precisely the brand of personal charm and the entrepreneurial spirit expressed by Natural Gift and the most profound essence made by the strong sense of design that deconstructing reorganization of the trends for each season; clothing style is full of irony and it is against going with the stream; expression of character traits in the confidence, sexy with elegance; and combination of high-level tailoring. Natural Gift also represents a kind of lifestyle: the brand tells the language of young people that it is delighted with the completely free experimental material and shape. Natural Gift lives in the geographic boundaries of contemporary metropolis, drawing the stimulus inspiration from the music, technology, construction and many other fields to transform into the leather series that are full of designing content. Alexander Chu expects to turn his philosophical spirit in the clothing into a flag and interpret and introduce its sensual and unique style all over the world.

2015/2016秋冬海派时尚流行趋势

变 variety
[海派未来风格]
SHANGHAI FUTURISTIC STYLE

沈沉
Shen Chen

东华大学服装·艺术设计学院
纺织品艺术设计专业教研室主任
上海水舞深造文化传播有限公司创意总监
国家纺织产品开发基地·特聘评估专家

颓废中重生的新科技艺术

人们对雾霾的厌恶和对晴朗天空的期待，迈开了恢复我们所失去信仰的步伐，带领我们从废墟、荒芜、庸俗中重生，告别蛮荒。暗淡、深沉的色彩里透着中性色、小面积对比色的光芒，重拾价值——尊重传统经典又展望科技未来。

主动调整与对接人机关系，各种主张、主义融汇在合成面料中，实现了完美调和。熔融涂层和起皱发泡处理的粗糙质感通过细致规整和精美叠褶的人工印迹，呈现出简约实用的工艺潮流和超凡脱俗的艺术与设计互为的转基因。

鼎天时尚 沈沉《十字济慈》
266g/㎡，涤纶75%，维纶25%

鼎天时尚 沈沉《温故知新》
566g/㎡，羊毛45%，涤纶25%，
酯纤维20%，维纶10%，起皱

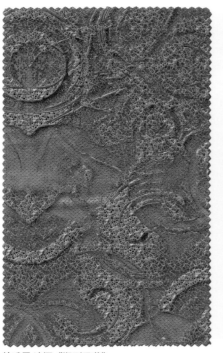

羊毛局 沈沉《塑型示范》
632g/㎡，维纶55%，涤纶45%，复合

鼎天时尚 沈沉《添那个玫瑰粉那个色》
651g/㎡，维纶35%，涤纶25%，棉7%，
聚酯5%，羊毛10%，复合

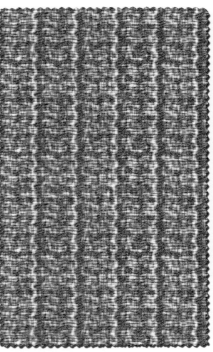

达利丝绸 沈沉《八卦水波形接回》
467g/㎡，维纶35%，涤纶25%，
棉15%，羊毛10%，聚酯5%，烂花

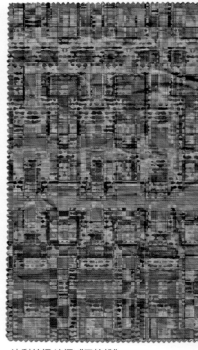

达利丝绸 沈沉《巴拉维》
367g/㎡，涤纶75%，维纶25%，梭织印花

2015/2016 AUTUMN/WINTER STYLE SHANGHAI FASHION TREND

[海派未来风格] SHANGHAI FUTURISTIC STYLE
variety 变

水舞深造 沈沉、龙帅雄《蝶翼》
866g/m²，聚酯纤维45%，粘胶30%，氨纶15%，羊毛10%，复合

REBIRTH OF NEW TECHNOLOGY ART IN DECADENCE

With the aversion to haze and expectation of clear sky, we take the pace that recovers our lost faith, guiding us to be reborn from ruin, desolation and vulgar, and bid farewell to the wild. Dark and deep colors reveal the neutral sense and contrasting color in small area, to regain the values – respect for traditional classic and prospect for future technology.

A perfect harmony is achieved through the initiative adjustment on the man-machine relationship and docking, and integration of various ideas and doctrines into the synthetic fabrics. Through the carefully structured and fine artificial pleats, the rough texture of foam melt coating and wrinkle treatment shows the simple and practical technology trends and transgenic vulgarity of art and design.

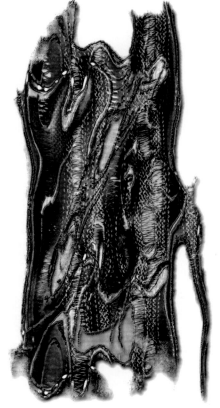

明阳集团 沈沉《千岛湖之水如铁》
794g/m²，棉%15，腈纶37%，羊毛48%，针织提花

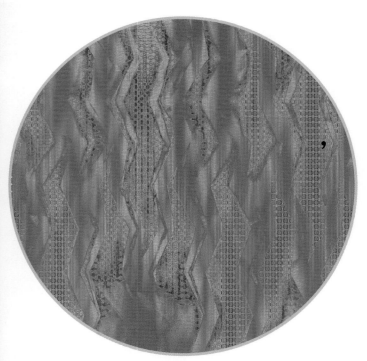

达利丝绸 沈沉《巴拉维汶川歌谣渐变》
433g/m²，羊毛48%，氨纶37%，棉15%，针织梭织复合花

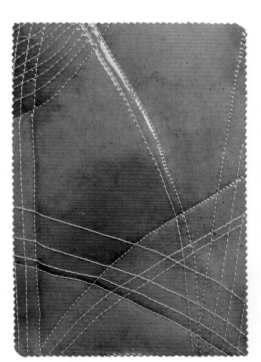

水舞深造 沈沉、廖畅《海蓝》
964g/m²，维纶75%，涤纶25%，复合

水舞深造丨沈沉丨张云《Fractal23》
297g/m²，超细涤纶63%，氨纶37%，梭织转印花

2015/2016秋冬海派时尚流行趋势

变 variety
[海派未来风格] SHANGHAI FUTURISTIC STYLE

倪明
Ni Ming

东华大学服装·艺术设计学院
纺织品艺术设计专业教师

"不变"中的变化

探究艺术创作创意产生的深层原因,揭示了"变"是设计的核心价值和意义,一成不变是腐朽落后的,而"变"又是于"不变"中得到发展和升华的。在以花卉为题材的图案中,写实风格的花卉能够具体而清晰的表现出一枝花的形态。同时,在自然界中存在着不计其数的个体集合,通过借助水墨技巧能够表现出抽象而含混的花团锦簇的形态。水墨技巧有章可循,而于运用之时又是变化万千。

对画理中的种种形式的运用,表面上看是造型与技术的问题,实质是设计理念和方法的问题。设计中不变的是对花卉的形态了然于心,变化的是对具象形态的抽象提炼和随心所欲的布置安排。

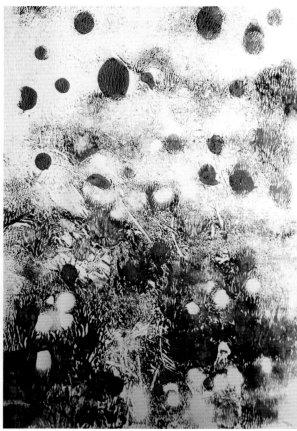

水墨意象花1/2/3

2015/2016 AUTUMN/WINTER STYLE SHANGHAI FASHION TREND

[海派未来风格]
SHANGHAI FUTURISTIC STYLE

variety 变

水墨意象花 4
水墨意象花 5/6

水墨意象花 8

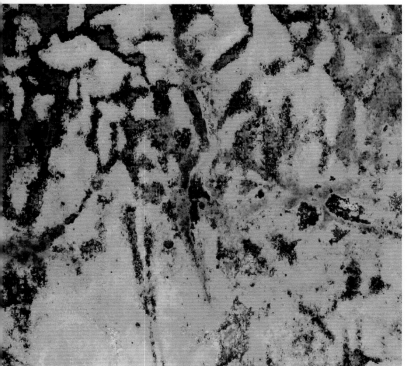
水墨意象花 7

THE CHANGE IN "INVARIABILITY"

When the underlying causes for the creativity of artistic works are explored, it is revealed that "change" is the core value and meaning of the design and it is decadent and backward with no change because the "change" is developed and sublimated in the "invariability". In the patterns with the floral theme, the specific and clear shape of a flower can be demonstrated in the flowers of realistic style. Meanwhile, there are countless collections of the individuals in the nature and the abstract and vague form of flowers can be expressed by means of the ink painting techniques that are based on some rule to follow with great changes in the use.

It is a problem of the style and technology on the surface, but it is a problem of the design concept and methods in fact based on the use of the various forms in the painting. The invariability in the design is to have a clear understanding of the shape of flowers and the change is to have an abstract refining and an arbitrary arrangement of the figurative forms.

2015/2016秋冬海派时尚流行趋势

变 variety

[海派未来风格] SHANGHAI FUTURISTIC STYLE

吴亮
Wu Liang

东华大学服装·艺术设计学院
视觉传达系副主任

畅想科技与太空

　　云科技为我们带来了难以想象的信息效率，三维打印与光场相机，使我们能够所想即所得。我们获得了古人难以想象的物理世界的自由，但人之间的巨大差异和隔阂是存在的，我们日益厌烦不变僵化的模式，这都让我们强烈向往着激情跨界，理性共生。精英们探求着莫测的蓝海，未来年轻人都想证明自己光热的存在。环境危机，恐惧宿命，大胆遥望未来的太空，对科技的期待是否能交换蜕变重生的到来。

李安琦

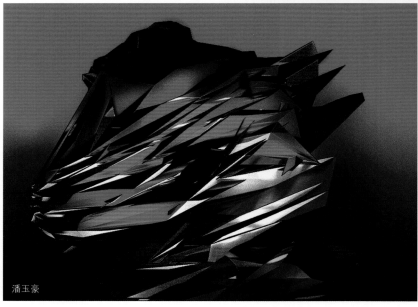

潘玉豪

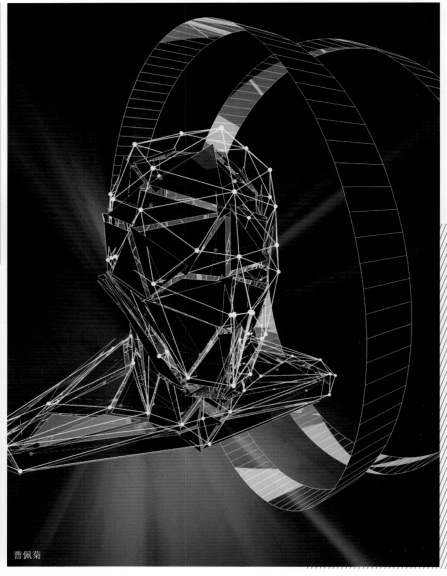

曹佩菊

2015/2016 AUTUMN/WINTER STYLE SHANGHAI FASHION TREND

[海派未来风格] 变
SHANGHAI FUTURISTIC STYLE

variety

THE PAMPER IMAGINATION OF THE TECHNOLOGY AND SPACE

The era of the cloud technology has brought the unimaginable information efficiency. You can have what you can image, such as the three-dimensional printing and optical field camera. We have gained the unimaginable freedom of the physical word in the old times but the huge difference and gap between people still exists. We are increasingly tired of the same rigid model, which makes us strongly yearn for the passionate cross-border and rational symbiosis. The elites are exploring the unpredictable blue ocean, the young people in the future all want to prove their fervent existence. Whether the environmental crisis, fate fear, bold overlooking into the future space and the expectation to the technology will exchange for the metamorphosis and rebirth or not.

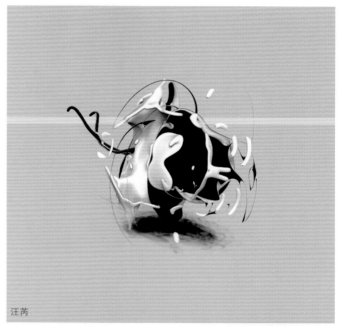

汪芮

唐迅韬

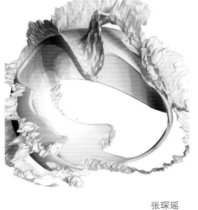

张琛瑶

周吉

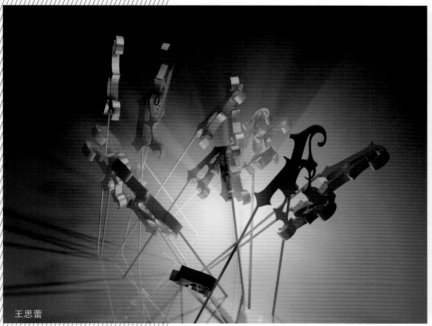

王思蕾

陈菲

2015/2016秋冬海派时尚流行趋势

变 variety
[海派未来风格]
SHANGHAI FUTURISTIC STYLE

田玉晶
Tian Yujing

东华大学服装·艺术设计学院
工业艺术设计系教师

戴伟豪 结构感帽子设计

未来配饰设计的建筑灵感

新生设计力量崛起，为我们带来了充满未来造型感的设计。注视、停止、思考，细节和时间中得出新的灵感，以全新眼光看世界，突破社会常规的设计，鼓励大家对于未来无论服装还是鞋履设计无限畅想，将那些不可能的变为可能。结合建筑的灵感来源，将建筑的细节融合到鞋履设计，未来感十足。

未来概念鞋履，犀利的结构感线条、充满想象空间的印纹、迷幻电子光感色调、独到的材质处理与精良的工艺相结合，宛若每只鞋子都充满灵性，称之为后现代艺术品也不为过，令人产生极度的占有欲望。这也必将引领鞋履设计发展的潮流。

戴伟豪 鞋履设计作品

 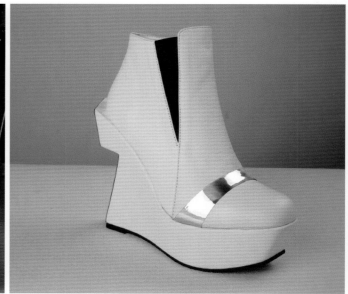

2015/2016 AUTUMN/WINTER STYLE SHANGHAI FASHION TREND

THE ARCHITECTURAL INSPIRATION OF THE FUTURE ACCESSORIES DESIGN

The emerging design strength gives us a sense of futuristic design style. Watching, stopping, and thinking. New inspiration derives from the details and time, to see the world with new vision, make the unconventional design, and encourage everyone to make the unlimited imagination regardless of future clothing or footwear design, transforming impossible into possible. Combined with the sources of architectural inspiration, the details of construction are integrated into footwear design, presenting the full sense of future.

Future concept footwear, sharp structure lines, fully imaginative printing, psychedelic colors with electronic light sensation, unique material processing and sophisticated technology are perfectly combined to present the full spirituality of each shoe, which can be called as post-modern art product to produce the extremely possessive desire. This will also lead to the development trend of footwear design.

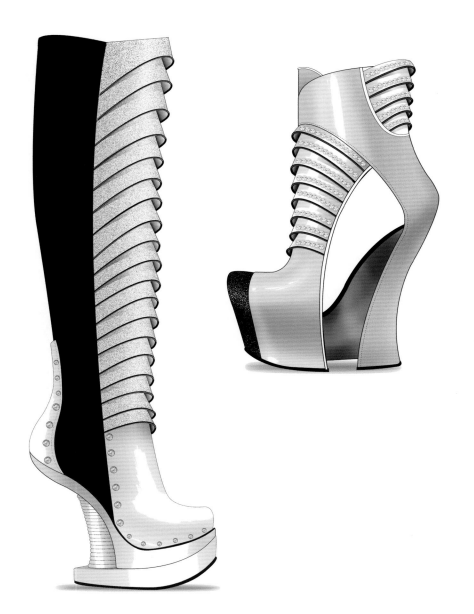

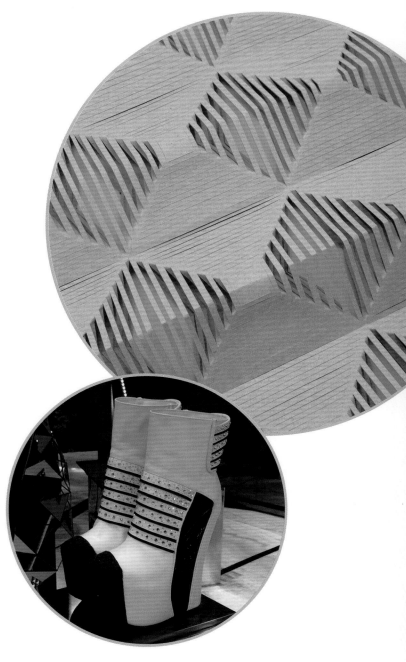

赵舟旭 鞋履设计作品

2015/2016秋冬海派时尚流行趋势

变 variety
[海派未来风格]
SHANGHAI FUTURISTIC STYLE

唐强
Tang Qiang

东华大学服装·艺术设计学院
艺术与科技专业副教授
东华古柏现代展示研究中心负责人

海派未来风格的变革思辨

作为推动世界转动的齿轮，变革无疑是任何事物、任何文明发展轴线上最为重要的一环，无论在现代与反现代，结构与解构中，在这一时间轴线上，我们可以逐渐理解现存和过去的其他文化的表达方式，其途径就是重构想象（reconstructive imagination）的训练。

在这种重构中，新的表达方式蜕变重生。当陷入现代主义的的解构与后现代整体疲软的状态中时，这一重构尤为重要。变革的脚步如浪潮般推动文明向前，在后国家时代文化边界逐渐消退的今天，如何保持文明本身特有的"自然属性"，又将文化本身融入具有国际化市场化的利益链条中，既是传统与未来视野间的一种吊诡（paradox）。海派未来风格就在这一时间节点成为这一主题不得不深入探讨的命题，如何保持海派风格本身的民族性文明特征，又不与其他在变革中被逐渐同化为利益驱使而失去其文化边界的风格沉瀣一气，关于海派风格的未来趋势的探讨就处于这样一种命题中，只有在这种文化的大命题下，关于空间视觉的可变性方式或者说未来主义的视觉陈列对于海派未来风格的探讨与研究才有其存在的意义，我们也将在这一大命题中从关于未来与传统的吊诡中看到变革的力量，同时发现这一力量将如何驱使海派风格的未来趋势与走向。

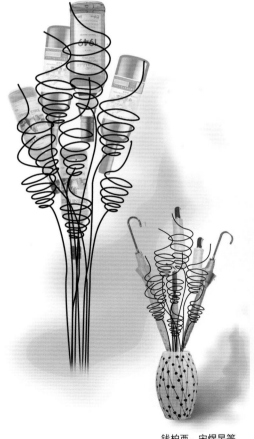

钱柏西、宋煜旻等

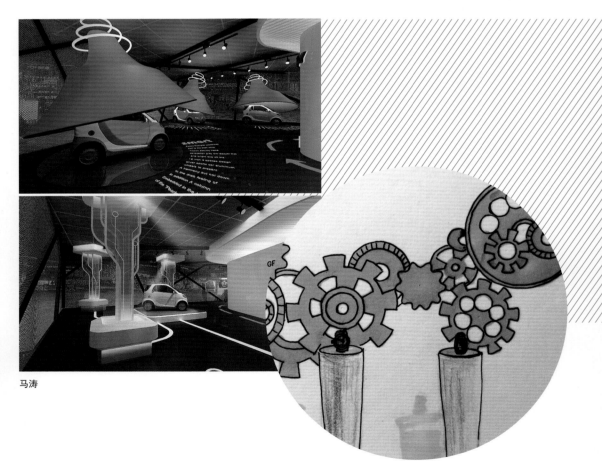

马涛

2015/2016 AUTUMN/WINTER STYLE SHANGHAI FASHION TREND

[海派未来风格] 变
SHANGHAI FUTURISTIC STYLE

variety

童劫

ANALYZE THE REVOLUTION OF NEW SHANGHAI STYLE

Being the motivation of promotion, whether in modernism or anti-modernism, construction or deconstruction, revolution is definitely the most important element in any cultural development. Under this circumstance can we learn other expressions of the existing and past culture by the training of reconstructive imagination.

There comes a new way of express in the whole process. When fall into a status that deconstruction of modernism and post-modernity are fatigued and weak, the reconstruction is particularly important. The progress of civilization based on revolution, how to keep the "natural quality" when the boundary of nations becomes indistinct, and incorporate culture itself into the international benefit chain, and that means the paradox of tradition coexist with modernity. The new Shanghai style turned into a proposition which deserves deep exploration. How to maintain the national character in Shanghai style in the meantime differentiate it from the style which lose its cultural boundary by assimilate in the revolution. The exploration of the new Shanghai style is in this proposition, and only under this circumstance, can the research of the variable ways of spatial visualization or visual merchandising has the meaning of existence. The power of revolution can be seen through the paradox of tradition coexist with modernity, meanwhile we will discover how it impel the trend of Shanghai style and development.

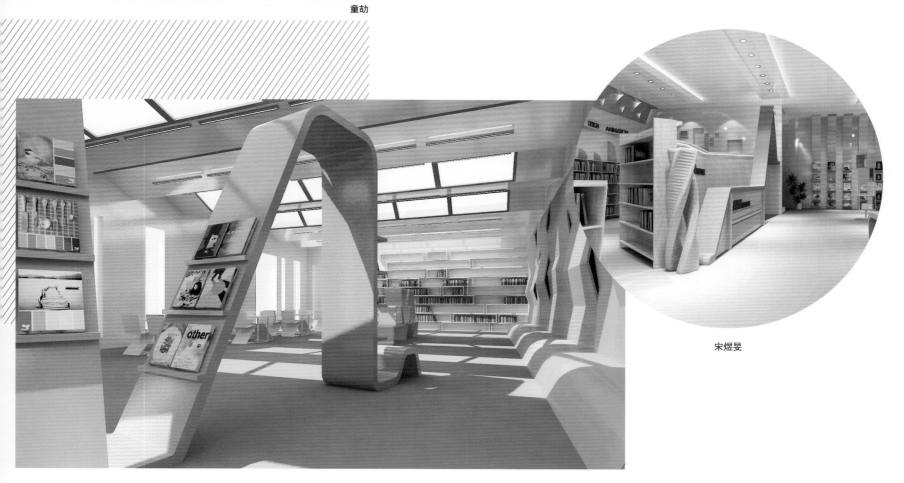

宋煜旻

鸣谢 ACKNOWLEDGEMENT

九木传盛品牌推进机构
上海标顶服饰有限公司
老凤祥名师设计中心
帽仕汇
红谷皮具有限公司
上海时湾艺术设计有限公司
上海洋滔品牌策划有限公司
达利丝绸（浙江）有限公司
上海水舞深造文化传播有限公司ZOOUZA 品牌
上海鼎天时尚科技股份有限公司
国际羊毛局（The Woolmark Company）
上海弄影时装有限公司
尚霞（上海）服饰文化传播有限公司
跨文化革新设计机构-字研所SHTYPE
德国FTA建筑设计有限公司
上海明阳集团
HELEN LEE品牌
时浪SNAP设计工作室
上海圆周率品牌传播机构
上海雅氏鞋业有限公司
卡拉羊
新秀集团有限公司
上海海琛国际贸易有限公司
LKK洛可可设计集团
ACIA亚洲创意产业联盟
MATCHBOX品牌
VIVINIKO品牌
东华大学-施华洛世奇创意设计中心
耐特利尔皮革时装有限公司
（以上排名不分先后，根据在本书中出现顺序排列）

图书在版编目（CIP）数据

海派时尚：2015/2016秋冬海派时尚流行趋势/
海派时尚流行趋势研究中心著.——上海：东华大学出版社，2014.5
ISBN 978-7-5669-0475-1
Ⅰ.①海⋯　Ⅱ.①海⋯　Ⅲ.①服饰-市场预测-上海市-
2015/2016　Ⅳ.①TS941.13
中国版本图书馆CIP数据核字(2014)第055480号

责任编辑　　杜亚玲

海派时尚：2015/2016秋冬海派时尚流行趋势
Haipai Shishang：2015/2016 Qiudong Haipai Shishang Liuxing Qushi
海派时尚流行趋势研究中心 著

出版：东华大学出版社(上海市延安西路1882号 邮政编码：200051)
出版社网址：http://www.dhupress.net
天猫旗舰店：http://dhdx.tmall.com
营销中心：021-62193056　62373056　62379558
印刷：上海中华商务联合印刷有限公司
开本：787mm×1092mm　1/8　　印张：17.5
字数：460千字
版次：2014年5月第1版
印次：2014年5月第1次印刷
书号：ISBN 978-7-5669-0475-1/TS•474
定价：380.00元